An Introduction to the Physics of High Energy Accelerators

WILEY SERIES IN BEAM PHYSICS AND ACCELERATOR TECHNOLOGY

Series Editor MEL MONTH

An Introduction to the Physics of High Energy Accelerators

D. A. EDWARDS
M. J. SYPHERS

SSC Laboratory
Dallas, Texas

WILEY-VCH

WILEY-VCH Verlag GmbH & Co. KGaA

To Helen and Ellen

A NOTE TO THE READER:
This book has been electronically reproduced from digital information stored at John Wiley & Sons, Inc. We are pleased that the use of this new technology will enable us to keep works of enduring scholarly value in print as long as there is a reasonable demand for them. The content of this book is identical to previous printings.

Library of Congress Card No.:
Applied for

British Library Cataloging-in-Publication Data:
A catalogue record for this book is available from the British Library

Bibliographic information published by
Die Deutsche Bibliothek
Die Deutsche Bibliothek lists this publication in the Deutsche Nationalbibliografie;
detailed bibliographic data is available in the Internet at <http://dnb.ddb.de>.

Printed on acid-free paper

ISBN-13: 978-0-471-55163-8
ISBN-10: 0-471-55163-5

Contents

Series Preface

More than 50 years ago the discovery and understanding of radiation led to the idea of a beam of elementary particles. Since then the development of processes for the collection, focusing, and acceleration of such beams has given rise to a growing number of facilities designed to produce a variety of particle beams for a multitude of purposes.

The higher particle energies achieved in accelerators have provided a major stimulus for research into the constituents and nature of matter. Since the 1930s new sciences, from atomic to nuclear to particle physics, have emerged concurrent with newly developed beam probes which allow research to proceed to smaller and smaller sizes, deeper into matter. Fueled by technological innovation and motivated by scientific curiosity, the increase in energy of particle accelerators has been about an order of magnitude every seven years. Present strong focusing synchrotrons can achieve TeV (10^{12} eV) energies—six orders of magnitude higher than the MeV (10^6 eV) energies achieved by the cyclotrons of 50 years ago.

Today, beams represent a novel state of matter of remarkable scope. During the past few decades, the energy of particle beams has been extended from keV to MeV to GeV to TeV; beam currents have gone from microamperes to milliamperes to amperes to megamperes; beam pulse durations have ranged from nanoseconds to continuous; beam lifetimes have been extended from microseconds to a week; species have multiplied from electron and proton beams to atomic, molecular, laser, and neutron beams, and then to pion, kaon, muon, neutrino, and antiproton beams; in addition, beams are bunched, squeezed, expanded, modulated, and chopped.

Beam physics is the study of particle and photon beams, their nature, their behavior, and their interactions, including the interaction of beams with matter, of beams with beams, and of particle beams with radiation. Evolving from concepts and ideas derived from classical mechanics, electromagnetism, statistical physics, and quantum physics, the study of beams is opening up a very rich field, with new effects being discovered and new types of beams with novel characteristics being realized. There are in fact a growing number of research areas in beam physics that need theoretical understanding and experimental verification, and these represent challenging opportunities for present and future generations of scientists and engineers.

The growth of beam physics is intimately tied to an expanding base of accelerator technology, including magnet systems, superconductivity, radiofrequency systems, particle sources, and others. The development of new types of particle beams is often tied to an extension of existing technology, for example, superconducting magnets for high energy accelerators, intense sources to produce very low emittance beams, and very high gradient accelerating fields to produce high energy beams in short distances. As a consequence, a substantial part of the research and development is directed toward advances in associated technologies.

The invention and continued development of particle accelerators and associated technologies have contributed profoundly to many subfields of pure and applied science and to our overall technological capabilities. In the fields of high energy and nuclear physics, a large complement of accelerator facilities exists around the world. In other areas of science, synchrotron light sources and accelerator driven pulsed neutron sources have opened up revolutionary new research opportunities in materials, chemistry, and biology. Hardly a field of science is not benefited. In industry and medicine there are thousands of accelerators in use in medical treatment, radiation sterilization, radiation processing, ion implantation, microchip production, etc. In defense, research is directed toward the use of particle beams in view of their very small transit times and potential ability to transmit large energy density over long distances.

The field of beam physics is still in its infancy, and many new developments have only begun to be applied to accelerators. The US Particle Accelerator school is dedicated to this new science. For more than a dozen years it has been organizing courses of study and publishing extensive compilations of pedagogical articles. The maturation of this rapidly growing field, however, requires the development of real text and reference books on important topics within it. For this reason, the school, in association with John Wiley & Sons, takes pride in presenting this new series on beam physics and accelerator technology. It is our hope that the school's continuing courses, enhanced by these new books, will help stimulate students, scientists, and engineers to contribute to the progress of this important and exciting field.

I would like to take this opportunity to extend to all the authors in this series our congratulations and gratitude for the enormous effort they are putting into this project and for the exceptional quality of the books they are writing.

MELVIN MONTH, DIRECTOR

US Particle Accelerator School

Preface

The traditional physics curriculum has at its core a series of courses in mechanics, electricity and magnetism, statistical physics, quantum physics, and other basics. This core may be supplemented by specialized courses the intent of which is to show how the principles come together to find application in a field of science or technology. The purpose of this textbook is to illustrate the application of the principles of physics in the field of high energy particle accelerators.

The first half of the book deals with the motion of a single particle under the influence of electric and magnetic fields. Much of the basic language of linear and circular accelerators is developed in this material. The principle of phase stability is introduced, and phase oscillations in linear accelerators and synchrotrons are studied. Next, the by now standard treatment of betatron oscillations is presented, followed by an excursion into nonlinear dynamics and its application to accelerators. Up to this point only one degree of freedom perturbations from the ideal motion will have been considered; we conclude the discussion of conservative single particle motion with a chapter devoted to coupling between the two transverse degrees of freedom.

In the first half of the book the particles are assumed neither to interact with each other nor to influence their environment. For the second half of the book, these restrictions are removed, and much of the discussion relates to intensity dependent effects—in particular, space charge and coherent instabilities. There is a wealth of phenomena that could be discussed under these headings; we will only be able to suggest the rich diversity by the few examples that space will permit. One chapter is devoted to common processes which may lead to the increase of transverse beam size in a synchrotron and provides some discussion about methods of its reduction. The dominant factor in the design of high energy electron accelerators is synchrotron radiation, and the basics of this process are the subject of the final chapter. We conclude with a bibliography and tables of parameters for a selection of accelerators which are used in the problems.

Homework problems are provided at the end of each chapter. These problems not only illustrate the material of the chapter but expand on it as well. Many of the problems are based on actual activities associated with the design and operation of existing accelerators. The problems are really an integral part of the text. In fact, most of the numerical content of our book is contained in them.

Throughout, we use the rationalized mks system of units, except where common usage dictates some other choice. The outstanding exception is in the specification of particle energy, where the electron-volt (eV) or its successive thousandfold multiples (keV, MeV, GeV, TeV) will be used. The electron-volt is the energy conveyed to a particle having one electronic unit of charge after it has been accelerated through an electrostatic potential diffference of one volt. So $1 \text{ eV} = 1.6 \times 10^{-19}$ J.

This book is the outgrowth of a graduate course in accelerator physics which we have taught over the past several years. In lecture note form this material has been used in standard quarter or semester courses at Northwestern University, the University of Texas at Austin, and at the University of Hamburg, Germany. The program of the US Particle Accelerator School has given us the opportunity to use the same material in intensive two-week courses at Cornell University, Harvard University, and Stanford University. In addition we have used our notes in informal courses at Fermilab, SSC Laboratory, and the International Centre for Theoretical Physics in Trieste, Italy. We have benefited greatly from the advice and comments of our students and colleagues.

We do not attempt to survey the entire field of accelerator physics. Rather, we have selected topics that are consistent with our experience and that we think are interesting. Since we have primarily worked with synchrotrons, much of the discussion is related to this type of accelerator.

A specialist in accelerator dynamics would likely use more sophisticated techniques than are used in much of the discussion. Though we do make use of the Hamiltonian formalism in some of the treatment of nonlinear motion, we feel that a more elementary approach is better suited for an introduction to the subject.

Our major published sources are referenced within the body of the text and in the bibliography at the end. In addition to the general expression of thanks to our students and colleagues above, we are especially grateful to our close associate Alex Chao for many insightful discussions, and to Mel Month for making possible our participation in the program of the US Particle Accelerator School and for encouraging the transformation of our rough notes into this book.

DON EDWARDS
MIKE SYPHERS

Cedar Hill, Texas

CHAPTER 1

Introduction

The use of macroscopic electromagnetic fields for the acceleration of charged particles dates from the mid-nineteenth century. The early x-ray tubes already contained in miniature the same systems—source, vacuum, power supply, accelerating structure—of later and much larger devices. In his discovery of the electron, J. J. Thompson toward the end of the century employed primitive particle accelerators of this sort. Today's cathode ray tube, which accelerates electrons to some tens of kilovolts, is the direct descendant of these devices.

Some decades elapsed before technology permitted the development of electrostatic potentials in the range of hundreds of kilovolts. Even as the progress in DC acceleration was accomplished, time varying electromagnetic fields were recognized as the route to still higher energies. The predecessor of the modern linear accelerator was developed first by Wideroe in 1928; the cyclotron was proposed in the following year by Lawrence, and acceleration was demonstrated in this first variety of circular accelerator in 1931. The development of the cyclotron proceeded rapidly, and by the end of the decade kinetic energy in excess of 10 MeV had been achieved.

The following decade saw the invention of the synchrotron, the synchrocyclotron, betatron, and the Alvarez linac. By 1950, protons and electrons had been accelerated to a kinetic energy of some 300 MeV. Yet higher energies were soon achieved, aided by the invention of the alternating gradient synchrotron. Figure 1.1 shows the dramatic increase in center of mass energy of the frontier accelerators with time.

Nuclear physics, and its descendant high energy physics, provided much of the motivation for the construction of particle accelerators until recently. But of late, the user community has expanded greatly beyond its traditional population as other applications of accelerators have been developed. Cases in point include synchrotron radiation sources and medical accelerators.

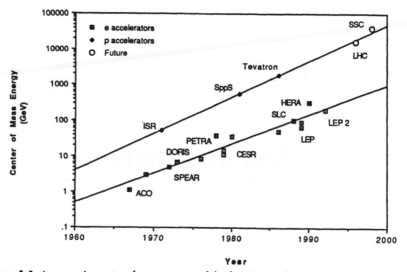

Figure 1.1. Increase in center of mass energy of the frontier accelerators with time. This figure is traditionally called a *Livingston plot*.

Figure 1.2. Aerial view of the Fermi National Accelerator Laboratory, in Batavia, Illinois. Photo courtesy of FNAL.

Figure 1.3. View of main tunnel at Fermilab showing the Tevatron (below) and original Main Ring synchrotrons. Photo courtesy of FNAL.

At the time of this writing, the highest energy particle accelerator in the world is the Tevatron at Fermilab, operating at a center of mass energy of 1.8 TeV as a proton-antiproton collider. (See Figures 1.2 and 1.3.) The highest energy electron devices are the SLAC Linear Collider and the LEP project at CERN; both develop 100 GeV electron-positron center of mass energies. A host of electron and positron storage rings on the GeV energy scale provide sources of synchrotron radiation for many scientific and industrial applications. For many years the venerable Harvard Cyclotron has been used in proton therapy; in recent months a 300 MeV proton synchrotron designed specifically for medical application has been brought into operation at the Loma Linda Medical Center in Southern California. An impressive array of ongoing projects is leading us toward even higher energies and broader applications.

1.1 PREREQUISITES

The bulk of the discussion in this text presumes that the reader has studied classical mechanics as well as electricity and magnetism at the level of completion of an undergraduate curriculum in physics or engineering. We assume some past exposure to special relativity. A brush with Hamilton's equations would be helpful, but is not essential. No background in accelerator or particle physics is needed. In the remainder of this section we sketch these essentials.

The equation of motion for a particle of charge e moving under the influence of electric and magnetic fields $\vec{E}$, $\vec{B}$ is

$$\frac{d\vec{p}}{dt} = e\left(\vec{E} + \vec{v} \times \vec{B}\right), \tag{1.1}$$

where $\vec{p} = \gamma m \vec{v}$ is the momentum, $\vec{v}$ is the velocity, and m is the invariant mass. In the expression for the momentum, γ is the Lorentz factor:

$$\gamma \equiv \frac{1}{\sqrt{1 - \dfrac{v^2}{c^2}}}, \tag{1.2}$$

$$v \equiv \sqrt{\vec{v} \cdot \vec{v}}. \tag{1.3}$$

The fields must satisfy Maxwell's equations. In a vacuum, Maxwell's equations in differential form are

$$\nabla \cdot \vec{E} = \frac{1}{\epsilon_0} \rho(\vec{r}, t), \tag{1.4}$$

$$\nabla \cdot \vec{B} = 0, \tag{1.5}$$

$$\nabla \times \vec{E} = -\frac{\partial \vec{B}}{\partial t}, \tag{1.6}$$

$$\nabla \times \vec{B} = \mu_0 \vec{j}(\vec{r}, t) + \frac{1}{c^2} \frac{\partial \vec{E}}{\partial t}, \tag{1.7}$$

where ρ and $\vec{j}$ are the charge and current densities respectively.

Frequently—particularly in simple geometries—the integral forms of Maxwell's equations are more convenient than the differential equations. These are

$$\int \vec{E} \cdot d\vec{S} = \frac{1}{\epsilon_0} \int \rho \, dV, \tag{1.8}$$

$$\int \vec{B} \cdot d\vec{S} = 0, \tag{1.9}$$

$$\oint \vec{E} \cdot d\vec{l} = -\int \dot{\vec{B}} \cdot d\vec{S}, \tag{1.10}$$

$$\oint \vec{B} \cdot d\vec{l} = \mu_0 \int \vec{j} \cdot d\vec{S} + \frac{1}{c^2} \int \dot{\vec{E}} \cdot d\vec{S}. \tag{1.11}$$

Here, the line, surface, and volume integrals are connected by the usual conventions. The physical interpretation is more transparent in the integral form. The first is Gauss's Law—a charge q is the source of q/ϵ_0 lines of $\vec{E}$. The second states that lines of $\vec{B}$ neither begin nor end; that is, there are no free magnetic poles. The third is Faraday's law of electromagnetic induction. The fourth is Ampere's law as modified by Maxwell to include the displacement current contribution.

We don't encounter any dielectric materials in this text, so there is no need to write down the equations in such media. But we will have occasion to use magnetic materials; then the last of Maxwell's equations becomes

$$\nabla \times \vec{H} = \vec{j}(\vec{r}, t) + \epsilon_0 \frac{\partial \vec{E}}{\partial t} \tag{1.12}$$

with $\vec{H}$ and $\vec{B}$ related through the magnetization $\vec{M}$ according to

$$\vec{B} = \mu_0 \vec{H} + \vec{M}. \tag{1.13}$$

In a simple conducting material, the current density is proportional to the electric field, where the constant of proportionality is the conductivity, σ: $\vec{j} = \sigma \vec{E}$. The permeability and permittivity of free space are related through $\mu_0 \epsilon_0 = 1/c^2$, and their values are $\mu_0 = 4\pi \cdot \times 10^{-7}$ henries per meter, $\epsilon_0 = 8.85 \times 10^{-12}$ farads per meter.

A slowly moving but accelerating charge will radiate power according to the Larmor formula

$$P = \frac{1}{6\pi\epsilon_0} \frac{e^2 a^2}{c^3}. \tag{1.14}$$

The angular distribution of the radiated power varies as $\sin^2 \theta$, where θ is the angle between the line joining the particle to the point of observation and the line along the direction of the acceleration of the particle.

Regarding relativity, we will occasionally make use of a Lorentz transformation. Suppose two inertial frames are moving relative to one another with speed v as in Figure 1.4. For clarity, the x and x' axes are shown with a separation; they actually are the same line. If their origins coincide at time $t = 0$, then the transformation equations are

$$x' = \gamma(x - vt) \tag{1.15}$$

$$y' = y, \tag{1.16}$$

$$z' = z, \tag{1.17}$$

$$t' = \gamma\left(t - \frac{xv}{c^2}\right). \tag{1.18}$$

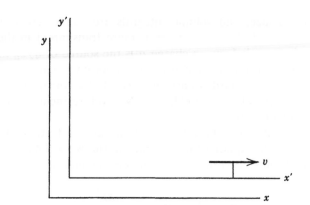

Figure 1.4. Inertial reference frames moving with respect to one another with relative speed v.

Two celebrated consequences of these transformations are time dilation and the Lorentz contraction. Time dilation can be obtained from the last equation. A clock in the primed frame located at $x = vt$ will show a time $t' = t/\gamma$. The Lorentz contraction follows from the first of the transformation equations. Suppose an object of length L' along the x'-axis is at rest in the primed frame with one end at the origin. At $t = 0$ the first transformation gives $L' = \gamma L$, so the length of the object in the unprimed frame is $L = L'/\gamma$.

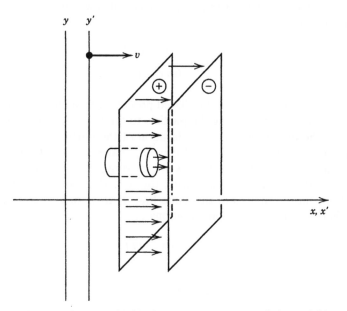

Figure 1.5. Application of Gauss's law to demonstrate invariance of electric field component parallel to direction of relative motion of two inertial frames.

The electromagnetic field transformations follow readily from the above. We will occasionally need to relate the fields in the rest frame of the beam to those in the laboratory frame. In the rest frame—the primed frame—there are only electric fields. To obtain the transformation of the electric field component in the direction of motion, suppose there is a large parallel plate capacitor at rest in the primed frame. Application of Gauss's law as in Figure 1.5 in both reference frames yields the result that $E_\parallel = E'_\parallel$, because the surface charge density is the same in both. The relationship between the transverse field components can be found by looking at the fields produced by a constant linear charge density λ' stretched along the x' axis. In the rest frame, $E'_\perp = \lambda'/(2\pi\epsilon_0 r')$. In the laboratory frame, the same expression holds, but without the primes. Due to the Lorentz contraction, the line density is larger in the lab frame by a factor of γ. The radial dimensions are unchanged by the transformation, so $E_\perp = \gamma E'_\perp$. Also, there is a transverse magnetic field in the lab frame, encircling the x-axis, because the moving line charge represents a current $I = \lambda v$. Using Ampere's law and comparing with the expression for the transverse electric field gives $B_\perp = \gamma E'_\perp v/c^2$.

1.2 USES OF ACCELERATORS

The construction of the early particle accelerators of the 1930s was motivated by nuclear physics. Now, some 60 years after the initial steps in the field, particle accelerators are found in a wide variety of applications, as illustrated by the by no means complete selection in the list below.

- Accelerators used for elementary particle physics. This category includes the highest energy per particle devices. The character of this field has changed as the notion of what "elementary" means has evolved.
- Accelerators for nuclear physics research. This now mature field, initiated in the 1930s, includes a broad spectrum of studies emphasizing energy precision, beam intensity, beam species, and polarized beams, ranging from traditional measurements of nuclear energy levels to the study of the quark-gluon plasma.
- Synchrotron radiation sources for a wide variety of applications of ultraviolet and x-ray beams in materials science. From a modest exploratory effort in 1952, this use has grown explosively. Users here far outnumber high energy physics experimenters.
- Accelerators for medical applications. These range from relatively low energy electron accelerators as x-ray sources to synchrotrons or linacs at the hundred MeV scale to provide hadron beams for radiation therapy.
- Ion beams for plasma heating in fusion reactor experiments.
- Electron linacs used for oil and natural gas exploration.

- Accelerators for food sterilization. This is more a potential than a real application at present; the countries that have need of the technology tend not to have the capital, and buyers in developed nations are suspicious of food that has been sterilized by these techniques.

Any of these topics could fill a textbook in itself. However, the same considerations concerning beam quality, focusing, stability, and so on are each common to many of them. Hence, we will spend the remainder of this section discussing two uses to serve as points of departure for the remainder of the text. The first is an example of a facility in which two beams are brought into collision in order to achieve high center of mass energy—a collider. The second example is of an accelerator whose use is derived from a single beam—a synchrotron radiation source.

1.2.1 Luminosity of a High Energy Collider

Because of the use of time varying fields to produce acceleration in high energy accelerators, these devices tend not to produce continuous particle beams, but rather beams that consist of a sequence of "bunches" of particles. In colliding beam physics, two such beams are brought into collision.

Suppose one bunch of particles moving in one direction collides head on with a bunch moving in the opposite direction. Let the bunches both have cross-sectional area A and both contain N particles. Any particle in one bunch will "see" a fraction of the area of the other bunch $N\sigma_{int}/A$ obscured by the interaction cross section σ_{int} (the total area of overlap of two colliding particles). This situation is suggested in Figure 1.6. For our purposes we allow the entire interaction cross section to be attributed to the particles in one beam encountering a point test particle in the other beam. The number

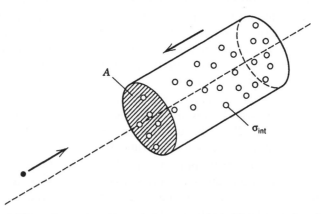

Figure 1.6. Collision of a single test particle from one beam with a particle bunch of the other beam.

of interactions per passage of two such bunches is then $N^2\sigma_{int}/A$. If the frequency of bunch collisions is f, then the interaction rate is

$$R = f\frac{N^2}{A}\sigma_{int}. \tag{1.19}$$

The luminosity, $\mathscr{L}$, is defined as the interaction rate per unit cross section:

$$\mathscr{L} = f\frac{N^2}{A}. \tag{1.20}$$

That is, the luminosity contains all the factors in the above expression that we think we can control and hence represents a figure of merit for a collider.

It would be surprising if beams came with bunches that were neat cylindrical uniformly populated volumes. Let us repeat the argument with a more realistic situation. Suppose that the particle distribution in the plane at right angles to the direction of motion is a "round Gaussian," or more properly, a Rayleigh distribution. As a function of radius, r, in the cylindrically symmetric distribution, the density function for the distribution is

$$dn(r) = \frac{N}{\sigma^2}e^{-r^2/2\sigma^2}r\,dr. \tag{1.21}$$

The contribution to the luminosity from two cylindrical shells at the same radius r and of the same thickness dr is

$$d\mathscr{L} = f\cdot\frac{dn}{2\pi r\,dr}\cdot dn. \tag{1.22}$$

Then, after adding up the contribution to the luminosity from shells of differing radius, we have

$$\mathscr{L} = f\frac{N^2}{4\pi\sigma^2}. \tag{1.23}$$

Luminosities are often expressed in cgs units, and the numbers look large and impressive. For instance, the CERN and Fermilab proton-antiproton colliders operate at a luminosity of more than 10^{30} cm^{-2} sec^{-1}. The total cross section is about 100 mb (1 millibarn $= 10^{-27}$ cm^2), so the interaction rate of 10^5/sec also looks large. But the cross sections of interest are much smaller—by a factor of about 10^8 for intermediate boson production. A more informative way to characterize collider performance is luminosity integrated over time and expressed in units of the relevant cross section. In that language, the 1988–89 collider run at Fermilab yielded an integrated luminosity greater than 10 pb^{-1} (1 picobarn $= 10^{-12}$ barns).

The point of the numbers above is to suggest that there is continual pressure to increase luminosity, particularly as the search for higher mass particles continues. The production cross section varies inversely as the square of the mass state, so in the case of the SSC (Superconducting Super Collider), the design goal is a luminosity of 10^{33} cm^{-2} sec^{-1}. Because of the difficulty of producing and storing large numbers of antiprotons, the SSC is designed as a two-ring proton-proton collider. Positrons are easier to come by than antiprotons, and today's electron-positron colliders routinely operate in the 10^{31} to 10^{32} luminosity range. The record for luminosity thus far is shared between the proton-proton intersecting storage rings at CERN and the Cornell Electron Storage Ring (electron-positron collider) at 1.3×10^{32}.

A glance at the luminosity formula reveals that to raise luminosity one must increase the collision frequency and bunch intensity and lower the beam cross sectional area. We will see that high bunch intensity moves toward collective single-bunch instabilities and beam-beam effect limitations, high collision frequency moves toward multibunch instabilities, and low transverse beam size places demands on beam sources and focusing systems.

1.2.2 Synchrotron Radiation Sources

Electrons circulating in a synchrotron are centripetally accelerated and so radiate just like electrons in an antenna. This process of *synchrotron radiation* is the topic of Chapter 8; we will just note here that the classical antenna formula is enhanced by a factor of γ^4 for relativistic motion characterized by a Lorentz factor γ. Synchrotron radiation was an irritant in early electron synchrotrons and storage rings because of the scale and cost implications for the radiofrequency acceleration system. But it was not long before it was realized that the synchrotron radiation was a valuable product in itself for research requiring intense, bright sources of ultraviolet light and x-rays.

The first *light sources* were synchrotrons designed for high energy physics research, and the light beams were generated by the main bending magnets. As demand for these facilities grew, storage synchrotrons designed expressly as light sources were constructed which incorporated special magnetic devices for enhancing and tailoring the radiation characteristics for this user community. For this discussion, we will assume that the radiation is produced by the particular device called an undulator. In contrast to bending magnets, an undulator produces a line spectrum (see Problems in Chapter 8); here we assume a single photon energy w.

If ΔE is the energy lost by one electron as it passes through the undulator, then the number of photons emitted from the undulator per second, the *flux*, can be written as

$$\mathcal{F} = \left(\frac{\Delta E}{w} \right) f_0 N_{\text{tot}}, \tag{1.24}$$

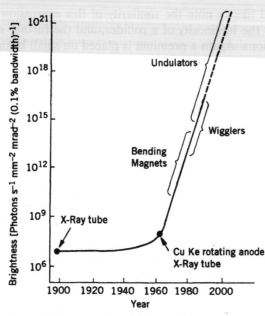

Figure 1.7. Development of source brightness. Reprinted from H. Winick, IEEE Conf. Proc. CH2669-0, 1989, with permission.

where f_0 is the revolution frequency of particles circulating in the synchrotron and N_{tot} is the total number of particles stored in the synchrotron. The quantity in parentheses is the number of photons emitted by one electron in a single passage through the undulator.

The figure of merit analogous to the luminosity of a collider is, in this case, the brightness, that is, the maximum value of the flux per unit area per unit solid angle. This phase space density is a conserved quantity throughout an optical system, and hence may be evaluated at the source. If we assume Gaussian distributions of particle position and direction transverse to the beam center, then the flux per unit phase space area has the form

$$\frac{d^4 \mathscr{F}}{dx\, dy\, d\theta\, d\psi} = \frac{\mathscr{F}}{(2\pi)^2 \sigma_x \sigma_y \sigma_\theta \sigma_\psi} e^{-(x^2/2\sigma_x^2 + y^2/2\sigma_y^2 + \cdots)}. \tag{1.25}$$

Implicit here is the assertion that the photons are emitted in the direction of motion of the electrons; we will discuss this point in Chapter 8.

The maximum value is obtained when all the transverse displacements are set to zero, and so the brightness is

$$\mathscr{B} = \frac{N_{\text{tot}}}{(2\pi)^2 \sigma_x \sigma_y \sigma_\theta \sigma_\psi} \left(\frac{\Delta E}{w}\right) f_0. \tag{1.26}$$

Here, we would like to note the similarity of this expression for the brightness to that for the luminosity of a collider, and the further similarity of the design implications. Again a premium is placed on small beam size, but here small beam divergence is also important. For high brightness, the beam current and energy loss through the undulator should be high.

Figure 1.7 illustrates the spectacular growth in brightness associated with the development of synchrotron radiation sources.[1]

PROBLEMS

1. The universe contains a superb accelerator. Some cosmic ray protons enter the top of the atmosphere with an energy of 1 joule or more. Calculate the difference in speed between the speed of light and a proton with an energy of 1 joule.

2. Derive the relativistic formula for addition of velocity. In the primed frame, a particle moves according to $x' = u't'$. Find the speed u in the unprimed frame.

3. The Stanford linac accelerates electrons to 50 GeV in a distance of 2 miles at a constant rate of increase of energy with distance. From the point of view of an observer riding precariously on an electron, how long would this journey last?

4. Starting from the definition of the work $dE = \vec{F} \cdot d\vec{s}$ done by a force $\vec{F}$ acting through a distance $d\vec{s}$ show that the energy gain ΔE of a particle of mass m associated with a change in Lorentz factor $\Delta\gamma$ is

$$\Delta E = (\Delta\gamma)mc^2.$$

If we define the rest energy $E_0 \equiv mc^2$, then the total energy (rest energy plus kinetic energy) is $E = \gamma mc^2$. Thus, the Lorentz factor γ is the ratio of the particle's total energy to its rest energy:

$$\gamma = \frac{E}{E_0}.$$

5. Using the result of the preceding problem, show that

$$E^2 = p^2c^2 + m^2c^4.$$

6. In the next chapter, we will make use of the relationship between fractional energy difference and fractional momentum difference. Show

[1]H. Winick, *Proc. 1989 Particle Accelerator Conf.*, IEEE 89CH2669-0, page 7.

that

$$\frac{dE}{E} = \left(\frac{v}{c}\right)^2 \frac{dp}{p}.$$

7. In the text, simple charge distributions were used to illustrate the transformation of electric fields from one frame to another moving with relative velocity v along their common x and x' axes. The same approach can be used to find the transformations for magnetic fields.

   (a) Suppose that a line current is directed along the positive x'-axis in the "primed" frame. In order that it produce a magnetic field only, assume that a positive line charge density λ' moves with speed u' in the positive x' direction and that a negative line charge density $-\lambda'$ is at rest. Then there is a magnetic field in the primed frame transverse to the x'-axis of magnitude

$$B'_\perp = \frac{\mu_0 \lambda' u'}{2\pi r'},$$

   where r' is the distance from the x' axis to the field point.

   (b) By transforming the charge distributions and their velocities to the laboratory frame, show that the magnetic and electric fields in the latter frame are given by

$$B_\perp = \gamma(v) B'_\perp \qquad E_\perp = \gamma(v) v B'_\perp .$$

   Remember that the positive line charge density λ' is not at rest in the primed frame, so its line density in the laboratory frame is not to be found just by dividing by the Lorentz factor relating the primed and unprimed frames.

8. No mention was made of gravity as a force to be taken into account here. Make an order of magnitude estimate to confirm that gravity can be neglected. The earth's magnetic field is approximately 1 gauss or 10^{-4} tesla—small compared with the various bending and focusing fields that we will consider in this book. Calculate the speed that a proton would have in order that the magnetic and gravitation forces would be equal in a field of 1 gauss. Calculate the corresponding kinetic energy.

9. A famous theorem in electrostatics states that the average of the electrostatic potential over the surface of a sphere containing no charges is equal to the value of the electrostatic potential at the center of the sphere. Prove this theorem.

10. The two-dimensional analog of the theorem in the preceding problem can be used to solve a variety of electromagnetic boundary value prob-

lems with modest computing resources. Suppose that we have a two-dimensional electrostatic field that varies with the coordinates x and y but not with z. Then the average of the electrostatic potential, $V(x, y)$, over the surface of a cylinder with axis parallel to the z-axis is the same as the value of V on the axis of the cylinder.

(a) Use this property of the electrostatic field to find the potential distribution within a square. On the boundaries of the square, the potentials will be specified. The potential distribution will be approximated by a grid. Each side of the square will be divided into 11 points, including the corners. Inside of the square there will be therefore $9 \times 9 = 81$ points. The potentials on the boundary points are selected by you; the problem is to determine the potentials within, using the averaging theorem.

(b) Find a small computer, and set up the problem. The algorithm goes as follows. You specify the potentials at the boundary of the square. As an initial guess, set all potentials at grid points within the square to zero. Then starting at, for example, the upper left hand corner interior point, find a new value at that point by averaging the potential values at the four neighboring grid points. The boundary values will begin to enter the interior. Proceed by rows or by columns, and observe the boundary potentials flow into the grid points within the square. A number of iterations are required; eventually you will see the potential distribution stabilize.

(c) Start with a problem whose answer you know. For instance, set the top edge of the square to 100 (volts), the bottom line to 0, and both edges to $0, 10, 20, \ldots, 100$ from bottom to top. Then set the top to 100, the left side to 20, the right side to 80, and keep the bottom edge at zero. Run, and sketch in the equipotentials and field lines in the result.

11. A simple bending magnet is made as shown in the figure below. N turns of conductor carrying current I are wound about each pole of the iron magnet. The poles are separated by a distance h. Assuming the permeability of the iron to be infinite, show that the field in the gap of the

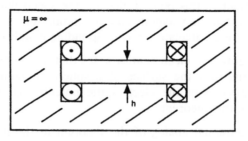

Figure for Problem 11.

magnet is given by

$$B = \frac{2\mu_0 NI}{h}.$$

12. A quadrupole magnet can be constructed as shown in the figure below. N turns of conductor carrying current I are wound about each pole of the iron magnet. The magnetic field has no z-component and is independent of z. Within the current free interior, the magnetic field can be expressed in terms of a magnetic scalar potential: $\vec{B} = \nabla \Phi_m$.

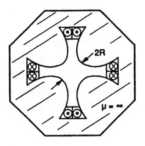

Figure for Problem 12.

(a) For the quadrupole magnet, where $\partial B_x/\partial y = \partial B_y/\partial x = B' = $ constant show that curves of constant Φ_m are hyperbolae.

(b) Each pole face of the quadrupole magnet is an "equipotential surface" given by $\Phi_m = B'xy = $ constant. If the distance between opposite pole faces is $2R$, and if the permeability of the iron is assumed to be infinite, show that the "gradient," B', of the magnetic field along the horizontal and vertical axes is given by

$$B' = \frac{2\mu_0 NI}{R^2}.$$

13. Generalize the previous problem to that of a $2n$-pole magnet:

(a) Show that the magnetic scalar potential satisfies Laplace's equation $\nabla^2 \Phi_m = 0$ and that its solution may be written as

$$\Phi_m = Cr^n \sin n\phi,$$

where C is a constant.

(b) Show that the magnetic field of the $2n$-pole magnet may be written in cylindrical coordinates as

$$B_r = Cnr^{n-1} \sin n\phi$$
$$B_\phi = Cnr^{n-1} \cos n\phi.$$

The constant C can be interpreted as

$$C = \frac{1}{n!}\left(\frac{\partial^{(n-1)}B_\phi}{\partial r^{(n-1)}}\right)_{\phi=0}.$$

(c) Suppose there are N turns of conductor per pole, each carrying current I. If the pole faces are equipotential surfaces with minimum radius R at angles $\phi = k \cdot (\pi/2n)$ for $k = 1, 3, 5, \ldots, 4n - 1$, show that

$$\left(\frac{\partial^{(n-1)}B_\phi}{\partial r^{(n-1)}}\right)_{\phi=0} = \frac{\mu_0 n! N I}{R^n}.$$

14. Consider a dipole magnet made from the following arrangement of current sources: Two cylindrical uniform current distributions, one with current density $\vec{J} = J\hat{z}$ and the other with current density $\vec{J} = -J\hat{z}$ have their centers separated by a distance d, as shown in the figure below. The current in the central region is exactly canceled. Show that the magnetic field within this region is given by

$$\vec{B} = B_x\hat{x} + B_y\hat{y} = \frac{\mu_0 d}{2}J\hat{y}.$$

Figure for Problem 14.

15. In Equation 1.23, the bunches collided head-to-head. Suppose instead that they collide with a small *crossing angle* α between their directions of motion. Assume that the distribution function of each bunch is also Gaussian in the direction of motion, with standard deviation σ_z. Then, show that the expression for the luminosity becomes

$$\mathcal{L} = \frac{fN^2}{4\pi\sigma^2}\frac{1}{\sqrt{1 + \alpha^2(\sigma_z^2/4\sigma^2)}}.$$

16. Two unbunched beams, each with λ particles per unit length distributed as a round Gaussian with standard deviation σ, collide with crossing

angle α. Show that the luminosity is

$$\mathscr{L} = \frac{c\lambda^2}{\pi^{1/2}\sigma\alpha}$$

in the high energy limit.

17. High luminosity comes more easily in fixed target physics than in collider operation. Suppose a beam of 10^{11} protons per second is incident on a liquid hydrogen target one meter in length. Calculate the luminosity. The density of liquid hydrogen is 0.07 g/cm^3. Assume that the protons are sufficiently energetic that they are traveling very close to the speed of light.

CHAPTER 2

Acceleration and Phase Stability

The most common particle accelerator is the ubiquitous cathode ray tube. But according to present convention, the term *particle accelerator* is usually reserved for multicomponent devices achieving particle energies of the order of 1 MeV or more. The use of an electrostatic field to provide the accelerating mechanism, as in the cathode ray tube, can be extended to energies on the order of 10 MeV per unit electronic charge. To reach higher energies, some repetitive acceleration procedure is normally employed. By joining a number of accelerating structures in series, the particle energy will increase in proportion to the distance traveled. In order that these structures provide electromotive forces, time varying fields are, by implication, involved. With present technology, acceleration up to 100 MeV per meter can be attained.

In circular accelerators, economical use of a single accelerating station is achieved by bending the path of the particles with magnetic fields so multiple passes are made through the station. To date, the highest energy (man-made) particle accelerators are of this variety.

Within both the straight and circular varieties, there are many variants. Most of our examples will relate to the conventional linear accelerator using microwave power sources to drive a sequence of accelerating stations—the *linac*—or to the type of circular accelerator called the synchrotron. Both came into prominence as the principal approaches to high energy acceleration just after World War II. In the case of the linac, radar components could be easily adapted to build the first of the big linacs. In the case of the synchrotron, the invention of the phase focusing principal in 1945 coupled with significant government support for research in particle physics made possible the construction of a first generation of synchrotrons, which came into operation in about 1950.

2.1 ACCELERATION METHODS

Our limited goal in this part of the chapter is to provide an elementary picture of the mechanisms of acceleration on which to base the discussion of longitudinal stability in linear accelerators and synchrotrons. After some introductory remarks on static and time varying electromagnetic fields, we discuss at some length a simple resonant cavity which could be used in linacs or synchrotrons. Although the structures actually used in these devices are generally more complex, our simple cavity could be used and is adequate for the main purpose of this chapter, namely the discussion of the principle of phase stability. We do not explore the technology of accelerating structures; that is a rich subject in itself. But we do conclude with a few remarks on the sorts of accelerating structures that are employed in practice.

2.1.1 DC Accelerators

The first accelerator in the modern sense was the electrostatic accelerator that Cockcroft and Walton used to demonstrate the disintegration of lithium nuclei upon bombardment with protons. They achieved a potential of 120,000 volts using a power supply based on the principle of charging capacitors in

Figure 2.1. Cockcroft-Walton preaccelerator at the Fermi National Accelerator Laboratory. This device provides negative hydrogen ions at 750,000 eV kinetic energy for subsequent acceleration through the Fermilab facility. Photo courtesy Fermilab.

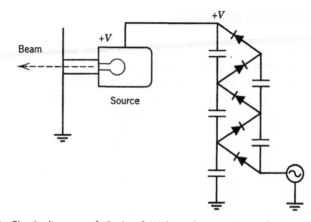

Figure 2.2. Circuit diagram of Cockcroft-Walton electrostatic accelerator. High voltage is produced by charging capacitors in parallel and discharging them in series.

parallel and discharging them in series. A quarter of a century later, the Cockcroft-Walton had become the standard preaccelerator for proton synchrotron facilities, operating at typically 750,000 volts. (See Figures 2.1 and 2.2.) At present, the Cockcroft-Waltons are being supplanted by the more compact and reliable radiofrequency quadrupoles.

The Van de Graaff generator produces high electrostatic voltages by the ingenious technique of spraying charge onto a moving nonconducting belt, which conveys the charges to an insulated terminal where the electron or ion source is located. (See Figure 2.3.) At the high voltage terminal, the charges are removed from the belt by the same multineedle type of comb that

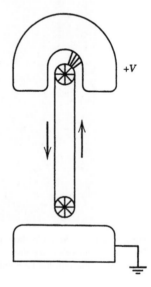

Figure 2.3. Circuit diagram of Van de Graaff electrostatic accelerator. High voltage is produced by collecting charge via a conveyor belt on an insulated terminal; charges to be accelerated are produced by a source located within the terminal.

sprayed them on in the first place. By pressurization of the tank containing the accelerator, potential differences up to about 15 MV can be achieved. In the tandem Van de Graaff, in effect the voltage is doubled. In the tandem, the ion source is located at ground potential and emits a beam of negative ions—for instance, H^-. At the high voltage terminal, the beam is passed through a foil which strips the electrons from the ions, and the protons continue to be accelerated, arriving at gound potential with twice the energy that would be associated with the electrostatic potential difference.

Electrostatic accelerators have the advantages of providing DC beams with narrow energy spread. Though the energy capability of this sort of device has slowly increased, ultimately it is limited by voltage breakdown, and for higher energies one is forced to turn to other approaches.

It might occur to one to cause the particle to pass through a DC potential difference repetitively as depicted in Figure 2.4. Here, by implication, the mechanism for causing the repetitive passages is a magnetic field to create a circular orbit. In order to assure that the field is nonzero only between the plates, the plates must extend to infinity. But then, the particle will be decelerated as it passes back through the plates half way around the orbit. In the situation of Figure 2.4 the plates are of finite extent and hence the electric field will be nonzero outside the capacitor; this "fringe field" will decelerate the particle as it approaches and departs the capacitor. No net acceleration will be achieved, as is of course consistent with the conservative nature of a static electric field. Therefore, as Faraday's law requires, we are compelled to consider electromagnetic fields which vary with time.

2.1.2 Time Varying Electromagnetic Fields

Production of an electromotive force in vacuum requires, according to Faraday's law, that the magnetic flux in some region of interest vary with time. There are many ways of employing Faraday's law to this end, each in a

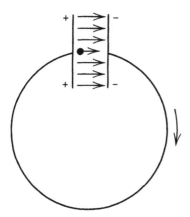

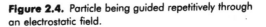

Figure 2.4. Particle being guided repetitively through an electrostatic field.

very real sense representing an invention. We listed a number of accelerator types in the Introduction. Here we will focus on one such, the betatron, which illustrates the principles in a transparent and elegant way as a precursor to the main theme of acceleration using resonant structures.

The betatron was the first circular accelerator to operate at constant orbit radius, and transverse particle oscillations in the neighborhood of the design trajectories of modern accelerators are still called betatron oscillations because they were studied in this case initially. In the betatron, a time varying magnetic field produces an electric field which accelerates the particles as we will see in the following discussion.

Suppose that there is a uniformly distributed magnetic flux with rate of change $\dot{\Phi}$ in a cylindrical region of space as depicted in Figure 2.5. Then at a radius r, there will be a tangentially directed electric field of magnitude $E = \dot{\Phi}/(2\pi r)$. If we now provide a magnetic field B at r, directed in the same sense as $\dot{\Phi}$, a particle can be made to follow a circular orbit of constant radius. From

$$\frac{1}{r} = \frac{eB}{p} \tag{2.1}$$

the condition for constant radius is

$$0 = -\frac{1}{r^2}\frac{dr}{dt} = \left(\frac{\dot{B}}{p} - \frac{B}{p^2}\dot{p}\right)e. \tag{2.2}$$

From the law of motion, $\dot{p}$ can be replaced by $e\dot{\Phi}/(2\pi r)$, and so we arrive at the condition

$$\dot{\Phi} = 2\pi r^2 \dot{B}. \tag{2.3}$$

That is, for motion at constant radius, the flux through the orbit must increase at twice the rate that it would if the field on the orbit were constant within the orbit.

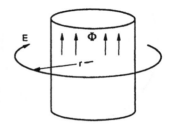

Figure 2.5. Principle of the betatron: accelerating electric field produced by changing magnetic flux through an orbit of radius r.

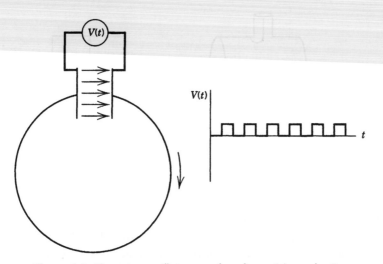

Figure 2.6. Elementary oscillatory waveform for particle acceleration.

The monotonic increase of flux to provide the accelerating field places a practical limit on the energy that can be achieved by the betatron principle. Nevertheless, the constant orbit radius of the betatron is attractive for high energy circular accelerators.

So we need to retain the feature of constant orbit radius while avoiding the monotonic field increase of the betatron; that is, we turn to oscillating fields. Suppose the capacitor of Figure 2.4 were energized only during the time when the particles pass through it. For instance, the waveform of Figure 2.6 would provide an electric field synchronized with particle passage, and the oscillatory character of the waveform indicates that the magnetic fields associated with charging the capacitor are also oscillatory in character.

One more step is necessary in order to complete our rudimentary picture of a high energy accelerator. Production of high voltages from square waves would be costly and technologically far from an optimum solution. (As will be seen later in this chapter, this waveform also fails to provide "phase focusing.") Typically, the fraction of the energy stored in the accelerating field that is transferred to the particles in each passage is small. Suppose a particle receives 1 MeV in passing the gap. If 10^{10} particles transit simultaneously as a bunch, then the total energy extracted from the accelerating device is only of the order of a millijoule. The energy stored in the electric field of the device typically will be several orders of magnitude larger. Therefore it makes sense to use a resonant structure for the accelerating device where the power delivered need only replace that extracted by the beam (and other losses).

We now have a model of a high energy accelerator. The particles encounter an electric field provided by one or more accelerating stations. Our

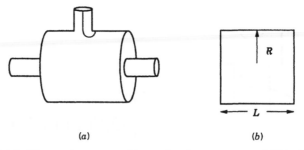

(a) (b)

Figure 2.7. (a) A pillbox cavity as a possible accelerating structure, and (b) its idealized model for the field calculation in the text.

model of a circular accelerator has one such station through which the particles pass repetitively as they are returned to it by a guiding magnetic field. A linear accelerator is composed of a sequence of such stations through which the particles make a single passage in achieving their final energy. The basic accelerating structure is often referred to as a *resonant cavity*; its elementary properties are described in the next section.

2.1.3 Resonant Cavities

A basic resonant cavity for use as an accelerating station might look like the cylindrical object depicted in Figure 2.7(a). A pillbox cavity has a couple of holes cut at either end to provide passage for the beam, and a third hole is provided on the cylinder surface for the coaxial drive. The current loop couples to the magnetic field of the cavity. So the idealized structure that we will look at is the closed cavity of Figure 2.7(b).

We want a mode with an accelerating field E_z, where z is in the direction of particle motion as indicated in the figure. Let us try to find a mode where this is the only component of the electric field, and where there is only one component of the magnetic field, B_θ. Maxwell's equations then reduce to

$$\frac{1}{r}\frac{\partial}{\partial r}(rB_\theta) = \frac{1}{c^2}\frac{\partial E_z}{\partial t}, \tag{2.4}$$

$$\frac{\partial E_z}{\partial r} = \frac{\partial B_\theta}{\partial t}. \tag{2.5}$$

After differentiating the first with respect to t and the second with respect to r, we can use the second to eliminate B_θ, with the result

$$\frac{\partial^2 E_z}{\partial r^2} + \frac{1}{r}\frac{\partial E_z}{\partial r} = \frac{1}{c^2}\frac{\partial^2 E_z}{\partial t^2}. \tag{2.6}$$

A mode with angular frequency ω will have a solution of the form

$$E_z(r,t) = E(r)e^{i\omega t}.$$ (2.7)

Therefore

$$E'' + \frac{E'}{r} + \left(\frac{\omega}{c}\right)^2 E = 0.$$ (2.8)

This is Bessel's equation of zero order, so the solution is

$$E(r) = E_0 J_0\left(\frac{\omega}{c}r\right).$$ (2.9)

At $r = R$, the field must vanish if the cavity material is a good conductor. The argument of the Bessel function must then be one of the zeros of J_0, and the lowest frequency mode will be associated with the first zero:

$$\frac{2\pi f}{c}R = 2.405.$$ (2.10)

With R at some reasonable scale, 30 cm say, then f is in the 400 MHz range, a perfectly appropriate frequency for radiofrequency power sources.

How long can the cavity be? Define the transit time factor T as the ratio of the energy actually given a particle that passes the cavity center at peak field to the energy that would be received if the field were constant with time at its peak value. That is, if the cavity length is L,

$$T = \frac{E_0\int_0^{L/2}\cos(\omega z/v)\,dz}{E_0(L/2)} = \frac{\sin u}{u}, \qquad u \equiv \frac{\omega L}{2v}.$$ (2.11)

If we ask that $T = 0.9$, for example, then for $v \approx c$, we have $u = 0.8$ and $L/R \approx \frac{2}{3}$.

The quality factor, or Q, of the cavity is defined as the ratio of the stored energy to the energy lost in one radian of an oscillation. When the electric field is a maximum, the magnetic field is zero, so we can calculate the stored energy from the electric field alone:

$$U = \tfrac{1}{2}\epsilon_0\int E^2\,dV$$ (2.12)

$$= \tfrac{1}{2}\epsilon_0 E_0^2 V J_1^2(2.405).$$ (2.13)

Here, we have used the definite integral

$$\int_0^1 J_0^2(\alpha x)\, x\, dx = \tfrac{1}{2}[J_0'(\alpha)]^2 \tag{2.14}$$

and the relation $J_0' = -J_1$. Since $J_1(2.405) = 0.52$, the stored energy is a little over 25% of the energy that would be associated with the field E_0 filling the entire volume V.

We next calculate the losses due to the ohmic heating in the cavity wall. Using Ampere's law, the surface current density is related to the magnetic field according to

$$J = \frac{1}{\mu_0} B_\theta, \tag{2.15}$$

where J is directed radially inward or outward on the end plates, and is parallel to the axes on the inner surface of the cylinder. If we write

$$B_\theta = B(r) e^{i\omega t}, \tag{2.16}$$

then, insofar as magnitudes are concerned

$$B(r) = \frac{E_0}{c} J_1\!\left(\frac{\omega}{c} r\right). \tag{2.17}$$

To get the loss rate, we integrate $\rho_s J^2/2$ over the surface, where ρ_s is the surface resistivity and the divisor 2 comes from averaging over the cycle. The surface resistivity is the bulk resistivity divided by the skin depth of the material at the frequency of interest. Surface resistivity therefore has units of ohms. Calling the loss rate P, we have

$$P = \tfrac{1}{2}\rho_s \left(\frac{E_0}{\mu_0 c}\right)^2 \left(2 \times 2\pi \int_0^R J_1^2\!\left(\frac{\omega r}{c}\right) r\, dr + 2\pi R L J_1^2\!\left(\frac{\omega R}{c}\right)\right) \tag{2.18}$$

$$= \tfrac{1}{2}\rho_s \frac{E_0^2}{Z_0^2} 2\pi R L \left(1 + \frac{R}{L}\right) J_1^2(2.405), \tag{2.19}$$

where Z_0 is the impedance of free space, $(\mu_0/\epsilon_0)^{1/2}$. The integral over the square of the Bessel function is

$$\int J_1^2(u)\, u\, du = \frac{u^2}{2} \{J_1^2(u) - J_0(u) J_2(u)\}, \tag{2.20}$$

but we note that the second term vanishes when evaluated at the first root of J_0. We can then estimate the Q from

$$Q \equiv \frac{\omega U}{P} \simeq \frac{2.405 \mu_0 c}{2\rho_s(1 + (R/L))}. \tag{2.21}$$

For copper surfaces in the 400 MHz region, this expression gives Q of order 10^4 for the geometry that we are using.

The Q is one figure of merit for a cavity. Another has to do with seeking high energy gain per unit loss. The ratio of the square of the energy gain to the loss rate has the dimensions of resistance; this quantity is called the shunt impedance of the cavity, R_s:

$$R_s \equiv \frac{(\text{energy gain per unit charge})^2}{P}. \tag{2.22}$$

The literature contains various different definitions of shunt impedance; the one above implies that the transit time factor is included in order to calculate the energy gain. For our case,

$$R_s = \frac{Z_0^2}{\pi \rho_s} \frac{L}{R} \frac{T^2}{(1 + (R/L)) J_1^2(2.405)}. \tag{2.23}$$

For the numbers that we have been using, R_s is about 7 MΩ.

The pillbox is easy to analyze, and could in fact be used in accelerators, such as linacs and electron synchrotrons, where variable frequency is not necessary. In practice, given the importance of RF power costs in such accelerators, much effort has gone into the evolution of a simple pillbox into an optimized shape.

For given stored energy, the energy gain may be increased somewhat through the addition of nose cones, shown in Figure 2.8(a). The losses occur predominantly on the cylindrical surface, and so higher Q can be obtained by spreading the currents out on a sphere-like surface. The resulting cavity would look something like the shape shown in Figure 2.8(b), and cavities of this style are used in the LEP electron-positron storage ring at CERN.

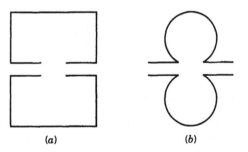

(a)   (b)

Figure 2.8. Variations of the pillbox cavity: (a) addition of nose cones; (b) spherical cavity to spread the surface currents.

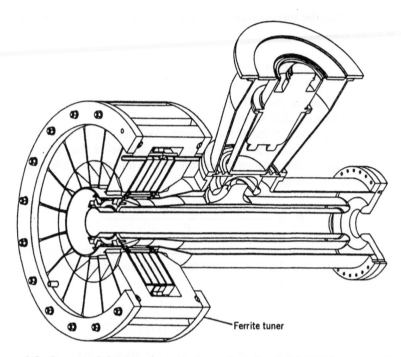

Ferrite tuner

Figure 2.9. Conceptual drawing of accelerating cavity to be used in SSC low energy booster synchrotron. The cavity uses ferrite tuners to adjust the cavity frequency from 48 to 60 MHz during the acceleration cycle.

Circular proton accelerators may require a significant frequency variation during the acceleration process. For instance, in the Fermilab booster synchrotron, the kinetic energy of protons is increased from 200 MeV to 8 GeV. The speed of the protons and hence the frequency of the accelerating system increases by a factor of 1.76 during this process. Suppose that our simple pillbox cavity were filled with a material of relative permeability $K_m \equiv \mu/\mu_0$. Setting aside for the moment the problem of getting the particle beam through this device, the resonant frequency of the cavity will have been reduced by a factor of $1/\sqrt{K_m}$. This can be easily seen by noting that Equation 2.4 will now contain a factor of K_m on the right hand side. Now we need a mechanism for varying K_m through the acceleration cycle. Figure 2.9 shows one approach to providing tunability while retaining the material-free passage for the beam.

2.1.4 Accelerating Structures

Though the discussion of the preceding section may be useful to illustrate the principles of acceleration in a large accelerator, in fact a single cavity is seldom sufficient to provide the requisite energy gain per turn in a synchrotron and surely not sufficient to produce the final energy of a linac. In a

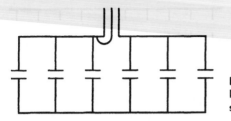
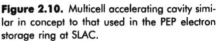

Figure 2.10. Multicell accelerating cavity similar in concept to that used in the PEP electron storage ring at SLAC.

variety of ways, arrays of cavities are grouped to form multicell accelerating structures. Often, it is convenient to drive such an array from a single source, as shown in Figure 2.10. Each individual cell if excited separately could oscillate at a common resonant frequency analogous to that of the pillbox. But when excited as a coupled system of five oscillators, the degeneracy is split and there will be five fundamental frequencies differing in the phase relationship of the fields from cell to cell. In the case of the structure in Figure 2.10, the cells are operated in the π-mode, where at a given instant in time the electric fields are in opposite directions in adjacent cells.

In a proton linac with its many cavities, fabrication methods and tolerances play an important role in the choice of structure. The proton linac LAMPF at Los Alamos National Laboratory employs the *side coupled cavity* structure shown in Figure 2.11(b). It can be thought of as a folded-up version of the $\pi/2$ mode arrangement indicated in Figure 2.11(a). In this mode, alternate cavities are unexcited in the lossless approximation. The structure at the left would have a low shunt impedance per unit length, because acceleration is provided by only half of the cells. But if the unexcited cells are moved off to the side, the acceleration efficiency of the π-mode case is recovered.

The above are examples of standing wave structures. This is a natural choice for an electron-positron storage ring, because it provides acceleration in both directions. In a proton linac, the speed of the particles changes throughout the accelerator, and one can design a series of structures with changing dimensions to match the change in energy.

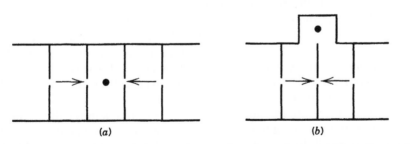

(a)   (b)

Figure 2.11. A cell structure (a) operating in a $\pi/2$ mode can be turned into a side-coupled cavity structure (b) such as that used in the LAMPF linac at LANL.

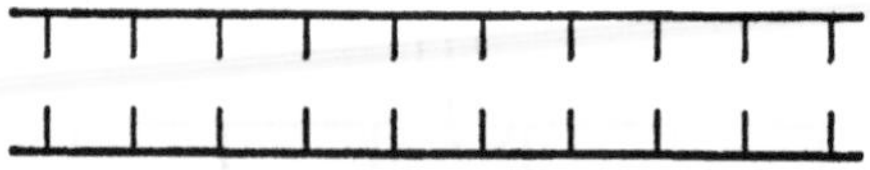

Figure 2.12. The disk-loaded wave guide
slows down the phase velocity of the accelerating wave to match the speed of the particles.

An electron linac is a different matter. An electron with a kinetic energy
of 3 MeV is already travelling at 99% of the speed of light. In this case, it is
natural to think of a uniform wave guide for the accelerating structure. But
the phase velocity in a hollow cylindrical wave guide is greater than c, so it is
necessary to slow the phase velocity down. This can be done in a variety of
ways, but the standard approach is to insert diaphragms into the guide. The
result is the standard disk-loaded wave guide depicted in Figure 2.12.
Electron linacs are usually designed for high frequency. The radar S-band
near 3 GHz was the choice for the Stanford two mile electron linac at SLAC.

Further discussion of accelerating structure technology would be out of
place in this text. A number of the references in the bibliography pursue this
subject in greater depth.

2.2 PHASE STABILITY

The rudimentary view of an accelerator that has emerged thus far is one in
which a particle traverses a number of accelerating devices—a possibly long
sequence of them in the case of a linear accelerator, and perhaps only one
for a circular accelerator, to which bending magnetic fields return the particle
from turn to turn. By implication, there is a special particle that adheres to
the perfect plan for the accelerator system. That is, there is a particle that at
each moment of time has exactly the right energy and the right time of
passage through the accelerating structure so that it receives exactly the right
increment of energy to stay in accord with the plan.

But we are not solely concerned with the history of an ideal particle;
rather, we must consider a distribution of particles differing in energy and in
the time between transits of the accelerating structure or structures. We thus
confront a classical stability question. Do particles initially nearby in energy
and transit time to those of the ideal particle remain nearby in this "phase
space" throughout the acceleration process? The principle that ensures that
this will be so is called the principle of *phase stability*.

We will find that there is a strong stability condition which dictates that
particles near the ideal particle will indeed remain nearby, and oscillate
about the ideal particle in energy–transit time space. These oscillations were
first analyzed for the device christened the synchrotron and so are called
synchrotron oscillations.

Of course, there are in reality three degrees of freedom for the motion of
the particle. We are going to describe the motion in the neighborhood of the
ideal particle in all three of them eventually. Variations in energy and transit
time are associated with one degree of freedom. Because transit time

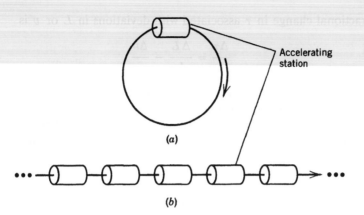

Figure 2.13. A sequence of accelerating stations as in a synchrotron (a) or in a linac (b).

differences are equivalent to position differences along the general direction of motion of the ideal particle, this is called the longitudinal degree of freedom. It is not yet apparent that it is possible to treat this degree of freedom independently of the other two, the so-called transverse degrees of freedom. We will see that the frequency of longitudinal oscillations is generally much less than that of transverse oscillations and so, to a reasonable approximation, they are decoupled.

2.2.1 Synchrotron Oscillations

As discussed previously, acceleration of particles to very high energies involves the use of high frequency resonant cavities for the production of accelerating fields. Suppose we have a sequence of such accelerating stations. A "station" may be regarded as either a cavity or system of cavities, such as in a synchrotron, or an individual accelerating gap within a cavity, such as in a linac. (See Figure 2.13.) Each station is excited by a source of radiofrequency power at angular frequency ω_{rf}. We assume that the system is designed so that the ideal particle arrives at each station at the same phase (modulo 2π) and receives the same increment of energy at each station.

The progress of the ideal particle through the accelerator is charted in the design of the device. In general, however, a particle will deviate from the design motion, and we wish to develop equations of motion that treat those deviations.

Let τ be the time interval between passages of two successive stations for the ideal particle. In terms of the spacing L between stations and the particle speed v,

$$\tau = L/v. \qquad (2.24)$$

The fractional change in τ associated with deviations in L or v is

$$\frac{\Delta\tau}{\tau} = \frac{\Delta L}{L} - \frac{\Delta v}{v}. \qquad (2.25)$$

That is, a particle moving with speed greater than that of the ideal particle will tend to take less time to pass between stations. But if its path length is larger, this deviation will tend to increase the transit time between stations.

The second term in Equation 2.25 can be expressed in terms of the fractional momentum deviation by

$$\frac{\Delta v}{v} = \frac{1}{\gamma^2}\left(\frac{\Delta p}{p}\right). \qquad (2.26)$$

In general, the first term might also be dependent upon the fractional momentum deviation. In a simple circular accelerator, for instance, one might expect the orbit circumference to be larger for a particle of momentum slightly above the momentum of the ideal particle. But for a linear accelerator the distance between stations is independent of the momentum deviation. Here, to cover these various cases, we will just introduce a parameter γ_t defined by

$$\frac{\Delta L}{L} = \frac{1}{\gamma_t^2}\left(\frac{\Delta p}{p}\right). \qquad (2.27)$$

The value of γ_t is determined by the type of device being studied as well as its particular design.

Our expression for the fractional change in τ is thus simplified by introducing the parameter

$$\eta \equiv \frac{1}{\gamma_t^2} - \frac{1}{\gamma^2}. \qquad (2.28)$$

Notice that at a particular energy $\gamma = \gamma_t$ the *slip factor* η changes sign. This is called the *transition energy*, whence the subscript t.

Now we can construct the equations governing the motion of a particle of arbitrary energy and phase. Suppose a particle arrives at the entrance to the nth accelerating station with energy E_n and phase ψ_n as depicted in Figure 2.14. (This could be the nth traversal of the same station for a circular

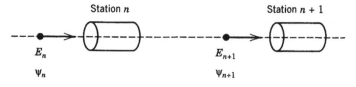

Figure 2.14. A particle enters the nth accelerating station with energy E_n and with phase ψ_n.

accelerator.) At the entrance to the $(n + 1)$st station the phase would be

$$\psi_{n+1} = \psi_n + \omega_{rf}(\tau + \Delta\tau)_{n+1}$$

$$= \psi_n + \omega_{rf}\tau_{n+1} + \omega_{rf}\tau_{n+1}\left(\frac{\Delta\tau}{\tau}\right)_{n+1}. \qquad (2.29)$$

Since the ideal, synchronous particle always arrives at the stations at the same phase (modulo 2π), it is convenient to switch to an angular variable which reflects this circumstance. We therefore define ϕ as

$$\phi_n \equiv \psi_n - \omega_{rf}T_n, \qquad (2.30)$$

where T_n is the time at entrance to the nth station. Then Equation 2.29 becomes

$$\phi_{n+1} + \omega_{rf}T_{n+1} = \phi_n + \omega_{rf}T_n + \omega_{rf}\tau_{n+1} + \omega_{rf}\tau_{n+1}\left(\frac{\Delta\tau}{\tau}\right)_{n+1}. \qquad (2.31)$$

But $T_n + \tau_{n+1} = T_{n+1}$ and so

$$\phi_{n+1} = \phi_n + \omega_{rf}\tau_{n+1}\left(\frac{\Delta\tau}{\tau}\right)_{n+1}$$

$$= \phi_n + \eta\omega_{rf}\tau_{n+1}\left(\frac{\Delta p}{p}\right)_{n+1}. \qquad (2.32)$$

For simplicity we will assume that $\omega_{rf}\tau$ is independent of n and so drop the subscript on τ in the following. For a circular accelerator the product $\omega_{rf}\tau$ is an integral multiple of 2π; this multiple is called the *harmonic number*. In a linear accelerator, the phase advance from cavity to cavity is set in the design of the structure; in Section 2.1.4, brief mention was made of the choices π and $\pi/2$.

We now have one of the difference equations of motion. The second difference equation treats the step in momentum, or more directly, the step in energy in passage through the accelerating station. If $(E_s)_n$ is the energy of the ideal, synchronous particle of charge e at the entrance to the nth station, then

$$(E_s)_{n+1} = (E_s)_n + eV \sin \phi_s, \qquad (2.33)$$

where V is the amplitude of the emf across the cavity gap and ϕ_s is the phase for arrival of the ideal particle, the so-called *synchronous phase*.

For a particle in general, the corresponding equation will be

$$E_{n+1} = E_n + eV \sin \phi_n, \qquad (2.34)$$

and so the difference in energy between the particle in question and the ideal particle, $\Delta E \equiv E - E_s$, must satisfy

$$\Delta E_{n+1} = \Delta E_n + eV(\sin \phi_n - \sin \phi_s). \qquad (2.35)$$

Since $E = \gamma mc^2$ and $p = \gamma mv$, one can show that

$$\frac{\Delta p}{p} = \frac{c^2}{v^2} \frac{\Delta E}{E}, \qquad (2.36)$$

and so the two difference equations for the motion of the particle with respect to that of the ideal particle are

$$\phi_{n+1} = \phi_n + \frac{\omega_{\mathrm{rf}} \tau \eta c^2}{v^2 E_s} \Delta E_{n+1}, \qquad (2.37)$$

$$\Delta E_{n+1} = \Delta E_n + eV(\sin \phi_n - \sin \phi_s). \qquad (2.38)$$

We can now address the question of whether or not a particle initially near to the ideal particle in energy and phase remains so as it proceeds through the accelerator. We will illustrate the principle for the case of a circular accelerator. For a linear accelerator, the argument is nearly identical; only a few words need be changed. The natural first step is to apply the difference equations to a variety of initial conditions. Figure 2.15 shows the result of a

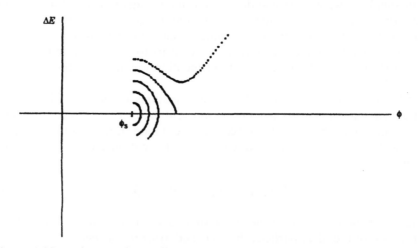

Figure 2.15. Application of the difference equations for synchrotron motion for five initial conditions. In each case, the starting value of the phase is equal to the synchronous phase.

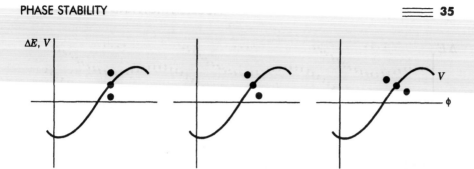

Figure 2.16. Three particles arrive at an accelerating station at the same phase: one with ideal energy, the other two with higher and lower energies. Stable excursions in energy and phase can exist.

few iterations of Equations 2.37 and 2.38 for a number of particles starting at ϕ_s on the first passage and differing in the initial values of ΔE. Observe that the particles closest to the ideal particle appear to remain close to the ideal particle as time develops. That is, the pattern of a stable oscillation is revealed. But the last particle, with the largest value of ΔE, is departing the neighborhood of the ideal particle. We may conclude that there is a finite region of stability.

Some aspects of the foregoing picture can be anticipated by a simple discussion. Suppose three particles traverse an accelerating station at the same time. One is at just the right energy for entry into this station and so has the so-called *synchronous* energy and receives just the right energy gain to continue the nominal acceleration process. The other two particles arrive at the same time, but have slightly higher and lower energies. Figure 2.16 depicts three successive traversals of the accelerating station of our circular accelerator example. On the first passage of the station, all three particles receive the same increment in energy. Suppose the slip factor η is negative. That is, speed changes are more important than path length changes. The higher energy particle thus arrives at the station the second time sooner than the ideal particle and hence receives less energy than the ideal particle receives. The lower energy particle acquires more energy than does the ideal particle. The energy difference has begun to shrink. The diminution in energy difference implies that the phase slip between the second and third passage will be less than that between the first and second passage. This reduction of both the phase change and the energy difference from turn to turn is the seed of stable oscillations. That is, the situation for stable excursions in energy and phase exists. This argument cannot identify the boundary between stability and instability; it only suggests that there is a region of stability.

We should point out that if we had carried out the foregoing discussion with the same initial conditions, including the same ϕ_s, but with a positive slip factor, we would have arrived at the conclusion that the motion is very likely unstable. We will discuss this point in more detail later in this section.

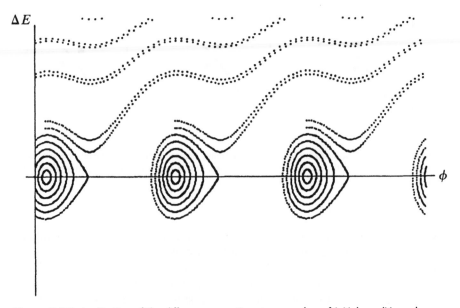

Figure 2.17. Application of the difference equations to a number of initial conditions demonstrates that regions of stable and unstable motion exist.

Now we return to the numerical solution of the difference equations to obtain a more detailed view of the trajectories in the ϕ-ΔE phase space . In Figure 2.17, the process initiated in Figure 2.15 is extended to a large number of turns and includes more cases. Several features are noteworthy. There is a well-defined boundary between stable and unstable motion. This boundary is called the *separatrix*. There are two qualitatively different points in phase space at which the particle undergoes no phase motion. The ideal particle occupies one of these points, called the stable fixed point, at $\phi = \phi_s$ and $\Delta E = 0$. The other lies on the separatrix and is called the unstable fixed point. The closer point separation in the neighborhood of the unstable fixed point indicates that particles move more slowly in this region; in particular, a particle on the separatrix and moving toward the fixed point would require infinitely many turns to reach it.

Figure 2.17 displays the circumstance that the harmonic number in a circular accelerator is generally larger than one and so there can be many stable fixed points distributed in azimuth, spaced in phase by 2π. The figure shows three such stable fixed points. The area in phase space within the separatrix is called a *bucket* in accelerator jargon, and so the figure depicts three stable buckets. All buckets need not be populated. For example, in the electron-positron storage ring (LEP) at CERN only four buckets are populated by each particle species. However, since the circumference of the LEP is 27 km and the bucket spacing is slightly less than a meter, just over a tenth of a percent of the buckets are occupied. Again, in the jargon, the collection of particles sharing a particular bucket is called a *bunch*.

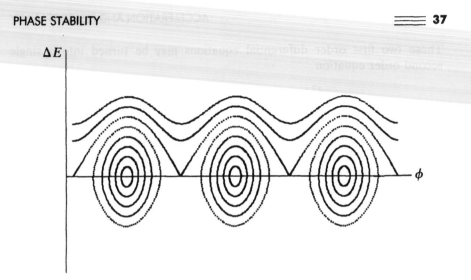

Figure 2.18. Application of the difference equations to a number of initial conditions for $\phi_s = 0$ or π. The regions within the separatrices are called stationary buckets.

Figure 2.17 shows a situation in which the ideal particle is undergoing acceleration. If the synchronous phase is 0 or π, we have the situation shown in Figure 2.18 in which the ideal particle is unaccelerated and the phase stable region is 2π in extent. These are called *stationary buckets*. For the case of the accelerating buckets of Figure 2.17 a particle outside the separatrix diverges in both energy and phase and ultimately will depart from the accelerator. But for this case, a particle outside the stationary bucket will only undulate in energy and may well remain within the accelerator indefinitely. It is still characterized as unstable, since it will wander progressively farther in phase from the ideal particle.

In a sense we have already solved the problem of phase and energy motion by numerical integration of the difference equations. We do not, however, have the convenience of a closed form analytical solution. A traditional analytical approach is to approximate the difference equations by differential equations. The legitimacy of treating phase and energy as continuous variables has already been suggested by the numerical treatment leading to Figures 2.15, 2.17, and 2.18 in that these dynamical variables change by rather small amounts from turn to turn. We may thus treat the turn number n as an independent variable and rewrite the difference equations as

$$\frac{d\phi}{dn} = \frac{\eta \omega_{rf} \tau c^2}{v^2 E_s} \Delta E, \tag{2.39}$$

$$\frac{d\Delta E}{dn} = eV(\sin \phi - \sin \phi_s). \tag{2.40}$$

These two first order differential equations may be turned into a single second order equation

$$\frac{d^2\phi}{dn^2} = \frac{\eta\omega_{rf}\tau eVc^2}{v^2E_s}(\sin\phi - \sin\phi_s),$$ (2.41)

provided that we assume a constant accelerating voltage and sufficiently small dE_s/dn. (We will relax this restriction in the following subsection.)

A first integral can be obtained by the standard prescription of multiplying by $d\phi/dn$ and integrating over n:

$$\int\frac{d^2\phi}{dn^2}\frac{d\phi}{dn}dn = \frac{\eta\omega_{rf}\tau eVc^2}{v^2E_s}\int(\sin\phi - \sin\phi_s)\frac{d\phi}{dn}dn$$ (2.42)

$$\Rightarrow \frac{1}{2}\left(\frac{d\phi}{dn}\right)^2 = -\frac{\eta\omega_{rf}\tau eVc^2}{v^2E_s}(\cos\phi + \phi\sin\phi_s) + \text{constant},$$

(2.43)

or,

$$\frac{1}{2}\left(\frac{d\phi}{dn}\right)^2 + \frac{\eta\omega_{rf}\tau eVc^2}{v^2E_s}(\cos\phi + \phi\sin\phi_s) = \text{constant}.$$ (2.44)

This is formally identical to the expression for the total "energy" $T + V = U$, where the first term on the left is the "kinetic energy" T, and the second term is the "potential energy" V. We can therefore make an energy level diagram where horizontal lines indicating the total "energy" of the particle are drawn in addition to the "potential energy" function V. The "potential energy" drawn in Figure 2.19 is for a nonzero value of $\sin\phi_s$. The

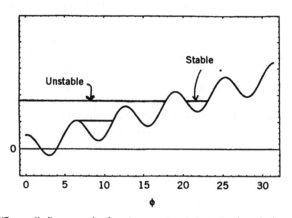

Figure 2.19. "Energy" diagram: the function $\cos\phi + \phi\sin\phi_s$ is plotted along with horizontal lines indicating levels of constant total "energy." The intersections of horizontal lines with the potential function indicate turning points of the motion, where $d\phi/dn = 0$.

intersection of the horizontal lines and the potential energy curve is a *turning point* where the "kinetic energy" term is zero. Of the three total "energies" shown, one depicts stable motion within a bucket, while another represents unstable motion unbounded to the left in the figure. The third represents a particle on the separatrix. The maximum extent in phase of stable motion on the separatrix is the subject of a problem at the end of the chapter. The unstable fixed point, at the limit of a converging sequence of turning points, is at $\phi = \pi - \phi_s$.

Combining Equations 2.39 and 2.44, we find the expression for contours describing particle motion in ϕ-ΔE phase space:

$$\Delta E^2 + \frac{2v^2 E_s eV}{\eta \omega_{rf} \tau c^2} (\cos \phi + \phi \sin \phi_s) = \text{constant}. \qquad (2.45)$$

The existence of these stable contours requires, strictly speaking, that the parameters such as E_s be constant. A number of trajectories in longitudinal phase space are shown in Figure 2.20 and indeed resemble the figures obtained by numerical integration of the difference equations.

We may now confidently use our analytical results to treat phase oscillations, having demonstrated the equivalence of the difference and differential approaches in our parameter range of interest. In order to obtain a simple expression for the frequency of phase oscillations it is, as usual, convenient to linearize the equation of motion. We assume that $\Delta\phi \equiv \phi - \phi_s$ is small and then the parenthetical expression on the right hand side of Equation 2.40 can be approximated by

$$\sin(\phi_s + \Delta\phi) - \sin\phi_s$$
$$= \cos\phi_s \sin\Delta\phi + \sin\phi_s \cos\Delta\phi - \sin\phi_s \qquad (2.46)$$
$$\approx \cos\phi_s \Delta\phi. \qquad (2.47)$$

Then

$$\frac{d^2\Delta\phi}{dn^2} + (2\pi\nu_s)^2 \Delta\phi = 0, \qquad (2.48)$$

where ν_s is the number of synchrotron oscillations per accelerating station passage, often referred to as the *synchrotron oscillation tune*. This quantity is given by

$$\nu_s = \sqrt{-\frac{\eta \omega_{rf} \tau c^2 eV \cos\phi_s}{4\pi^2 v^2 E_s}}, \qquad (2.49)$$

and the angular frequency of synchrotron oscillations is

$$\frac{2\pi\nu_s}{\tau} \equiv \Omega_s = \sqrt{-\frac{\eta \omega_{rf} c^2 eV \cos\phi_s}{\tau v^2 E_s}}. \qquad (2.50)$$

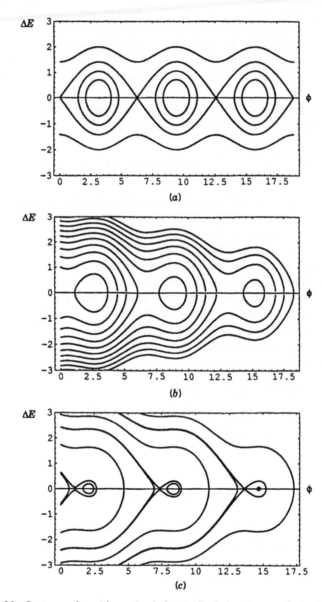

Figure 2.20. Contours of particle motion in longitudinal phase space obtained by solving the differential equations of motion. The curves shown are for $\eta > 0$ and for the cases (a) $\phi_s = \pi$, (b) $\phi_s = 5\pi / 6$, and (c) $\phi_s = 2\pi / 3$.

From this expression, we see that the correct choice for the synchronous phase depends upon the value of η as anticipated by our earlier discussion. For $\eta < 0$, that is, $\gamma < \gamma_t$, the motion is stable for $\cos \phi_s > 0$. However, for $\eta > 0$ $(\gamma > \gamma_t)$, the synchronous phase must be such that $\cos \phi_s < 0$. Circular accelerators that cross the transition energy during the acceleration cycle must perform a phase jump in the radiofrequency system at the appropriate time in order to maintain phase stability. Note that at the transition energy the synchrotron oscillation period becomes infinite and there is no phase focusing.

In the case of a linac, the slip factor η could be replaced by $-1/\gamma^2$, because particles of different momenta see essentially no difference in path length between accelerating stations. From Equation 2.49, one sees that as the particle energy increases, the synchrotron frequency approaches zero. As an example, for an energetic electron where $\gamma \gg 1$, the solution to the Equation 2.39 would look more like $\Delta\phi = $ constant: an electron starting out at a particular phase will remain at that phase.

The equation of motion for phase oscillation, Equation 2.41, is a nonlinear differential equation; the synchrotron tune, ν_s, just discussed is valid only for small oscillation amplitudes. On the separatrix, as already mentioned, the oscillation period is infinitely long; the synchrotron oscillation tune decreases monotonically with increasing amplitude (see Problem 2.11).

2.2.2 Adiabatic Damping and Longitudinal Emittance

In the preceding subsection, the change in the synchronous energy and other parameters was assumed to be sufficiently small over the period of interest so that their derivatives could be ignored. We now relax that restriction. We will find that if these parameters are permitted to change slowly over time, the concomittent changes in the oscillation amplitudes are given by simple expressions.

In order to permit analytical solution we will limit the following discussion to small oscillations about the synchronous phase. A more natural choice of independent variable for this purpose is time, t, rather than accelerating station passage number, n. We begin with Equations 2.33 and 2.34 recast as differential equations:

$$\frac{dE}{dn} = eV \sin \phi, \qquad (2.51)$$

$$\frac{dE_s}{dn} = eV \sin \phi_s. \qquad (2.52)$$

Noting that

$$\frac{d}{dn} = \frac{dt}{dn}\frac{d}{dt} = \tau\frac{d}{dt}, \qquad (2.53)$$

the foregoing become

$$\tau(E)\frac{dE}{dt} = eV \sin \phi, \tag{2.54}$$

$$\tau(E_s)\frac{dE_s}{dt} = eV \sin \phi_s. \tag{2.55}$$

Upon subtracting, we find

$$\tau(E)\frac{dE}{dt} - \tau(E_s)\frac{dE_s}{dt} = eV(\sin \phi - \sin \phi_s). \tag{2.56}$$

If we expand $\tau(E)$ about E_s,

$$\tau(E) = \tau(E_s) + \left(\frac{d\tau}{dE}\right)_{E_s}(E - E_s), \tag{2.57}$$

then to first order in ΔE, Equation 2.56 becomes

$$\frac{d}{dt}(\tau \Delta E) = eV(\sin \phi - \sin \phi_s). \tag{2.58}$$

The conversion of the phase equation to employ time as the independent variable is straightforward. Our pair of linearized differential equations is then

$$\frac{d\Delta\phi}{dt} = \frac{\eta \omega_{rf} c^2}{v^2 E_s} \Delta E \equiv \lambda \, \Delta E, \tag{2.59}$$

$$\frac{d}{dt}(\tau \Delta E) = \tau\left(\frac{eV \cos \phi_s}{\tau}\right)\Delta\phi \equiv \tau\mu \, \Delta\phi. \tag{2.60}$$

Combining these two equations and taking into account that the quantities λ and τ are now functions of time gives the second order differential equation

$$\frac{d^2}{dt^2}\Delta\phi - \frac{1}{\lambda/\tau}\frac{d}{dt}\left(\frac{\lambda}{\tau}\right)\frac{d}{dt}\Delta\phi + \Omega_s^2 \, \Delta\phi = 0, \tag{2.61}$$

where $\Omega_s = \sqrt{-\lambda\mu}$ is the angular frequency of synchrotron oscillations given in Equation 2.50. Note that since λ contains η while μ contains $\cos \phi_s$, then either $\lambda > 0$ and $\mu < 0$ or $\lambda < 0$ and $\mu > 0$.

A standard approach to solving a second order differential equation of this sort is to choose a trial solution of the form $\Delta\phi = uv$ and pick v such that the first derivative term is zero in the equation for u. Substituting, we find

$$v\frac{d^2u}{dt^2} + \left[2\frac{dv}{dt} - \frac{1}{\lambda/\tau}\frac{d}{dt}\left(\frac{\lambda}{\tau}\right)v\right]\frac{du}{dt}$$

$$+ \left[\frac{d^2v}{dt^2} - \frac{1}{\lambda/\tau}\frac{d}{dt}\left(\frac{\lambda}{\tau}\right)\frac{dv}{dt} + \Omega_s^2 v\right]u = 0, \qquad (2.62)$$

and so the indicated choice for v is to satisfy

$$2\frac{dv}{dt} = \frac{1}{\lambda/\tau}\frac{d}{dt}\left(\frac{\lambda}{\tau}\right)v. \qquad (2.63)$$

Therefore, we may choose $v = \sqrt{\pm\lambda/\tau}$, where the quantity under the radical has to be positive. Then, the differential equation for u becomes

$$\frac{d^2u}{dt^2} + \left\{\frac{1}{2(\lambda/\tau)}\frac{d^2}{dt^2}\left(\frac{\lambda}{\tau}\right) - \frac{3}{4}\frac{1}{(\lambda/\tau)^2}\left[\frac{d}{dt}\left(\frac{\lambda}{\tau}\right)\right]^2 + \Omega_s^2\right\}u = 0. \quad (2.64)$$

We next proceed to solve the differential equation for u using the method of integrated phase. That is, a solution is sought of the form

$$u = f(t)\exp\left\{i\int\omega\,dt\right\}, \qquad (2.65)$$

where f is a slowly varying function of time and ω^2 is the quantity in braces in Equation 2.64. Since, by our assumption, the first two quantities in the braces are small compared to Ω_s^2 we set ω in our trial solution equal to Ω_s. Then, the equation for f becomes

$$\ddot{f} + i\left(2\dot{f}\Omega_s + f\dot{\Omega}_s\right) = 0. \qquad (2.66)$$

Here, a dot refers to a derivative with respect to time. Since f is slowly varying, the first term is negligible compared to the second term, and so we have

$$f = \frac{1}{\sqrt{\Omega_s}}. \qquad (2.67)$$

Combining our results, we may write the solution for the phase oscillation in the form

$$\Delta\phi = A\sqrt{\frac{\pm\lambda}{\tau\Omega_s}}\cos\left(\int\Omega_s\,dt + \delta_1\right), \tag{2.68}$$

where A and δ_1 reflect the initial conditions and we have expressed the result in terms of real functions. Inserting the definitions of λ and Ω_s, we see that

$$\sqrt{\frac{\lambda}{\tau\Omega_s}} = \left[\pm\frac{\eta\omega_{rf}c^2}{v^2\tau E_s eV\cos\phi_s}\right]^{1/4}. \tag{2.69}$$

For instance, in the high energy limit of a circular accelerator where $\gamma \gg \gamma_t$, the phase oscillation amplitude varies inversely as the fourth root of the synchronous energy. This is an example of adiabatic damping.

We now consider energy oscillations. The differential equation for $\tau\,\Delta E$ is similar to that for $\Delta\phi$, namely,

$$\frac{d^2}{dt^2}(\tau\Delta E) - \frac{1}{\tau\mu}\frac{d}{dt}(\tau\mu)\frac{d}{dt}(\tau\Delta E) + \Omega_s^2(\tau\Delta E) = 0. \tag{2.70}$$

Proceeding exactly as in the case for $\Delta\phi$, we arrive at the solution

$$\Delta E = B\sqrt{\frac{\mp\mu}{\tau\Omega_s}}\cos\left(\int\Omega_s\,dt + \delta_2\right), \tag{2.71}$$

where

$$\sqrt{\frac{\mu}{\tau\Omega_s}} = \left[\mp\frac{v^2 E_s eV\cos\phi_s}{\eta\omega_{rf}\tau^3 c^2}\right]^{1/4}. \tag{2.72}$$

Looking again at our example of a particle with $\gamma \gg \gamma_t$, the oscillation amplitude in ΔE increases with synchronous energy, but the fractional energy difference $\Delta E/E_s$ will decrease.

Our discussion thus far has described the motion in a phase space in which the variables are ΔE and $\Delta\phi$. Is this the most natural choice of coordinates? Classical mechanics tells us that a simplification in the description of motion is gained by the use of canonically conjugate pairs. We can use our results to lead us to a familiar conjugate pair. From classical mechanics we know that for a particle undergoing periodic motion the area of its trajectory in the appropriate phase space is an adiabatic invariant. Observe that the product of the amplitudes of $\Delta\phi$ and ΔE as obtained above varies as $1/\tau$. That is, the area of the $\Delta\phi$-ΔE phase space ellipse is

$$\pi\,\Delta\hat{\phi}\,\Delta\hat{E} = \frac{\pi AB}{\tau}, \tag{2.73}$$

where a circumflex denotes an amplitude.

If, on the other hand, we use excursion of time of passage, Δt, rather than excursion of phase of passage, $\Delta\phi$, then $\Delta\hat{\phi} = \omega_{rf}\,\Delta\hat{t}$ and in ΔE-Δt coordinates the area becomes

$$\pi\,\Delta\hat{t}\,\Delta\hat{E} = \frac{\pi AB}{\omega_{rf}\tau}, \tag{2.74}$$

which is a constant. Of course E and t are well known to be a canonically conjugate pair.

The area in phase space which contains the particles of a bunch is termed the *longitudinal emittance*. It is to be hoped that this emittance is smaller than the area of the entire stable region. In ΔE-Δt coordinates, for a bucket with $\phi_s = 0$ or $180°$, the area of the bucket is

$$\mathscr{A} = \frac{16(v/c)}{\omega_{rf}}\sqrt{\frac{eV\cdot E_s}{2\pi h|\eta|}}. \tag{2.75}$$

As noted earlier, such a bucket is referred to as a *stationary* bucket, since the ideal particle is not accelerated in this case. This value of ϕ_s might be chosen, for example, at the injection level of a circular accelerator ($\phi_s = 0$ if below transition), or at the maximum energy ($\phi_s = 180°$ if above transition). In such a bucket, if $\Delta\hat{\phi}$ is the maximum extent of the presumed small oscillations of $\Delta\phi$, then the beam will have a longitudinal emittance

$$S = \frac{\pi(v/c)(\Delta\hat{\phi})^2}{\omega_{rf}}\sqrt{\frac{eV\cdot E_s}{2\pi h|\eta|}}. \tag{2.76}$$

When the synchronous phase is neither zero nor $180°$, the bucket area can be obtained numerically. Figure 2.21 shows how the bucket area varies with ϕ_s.

In these last few pages, we have made frequent use of a requirement that the parameters of the system vary sufficiently slowly that adiabaticity prevails. At transition, for example, our expressions would tell us that the synchrotron oscillation frequency goes to zero, the amplitude of the phase oscillation goes to zero, and the energy excursions go to infinity. This, of course, does not happen; many accelerators indeed cross transition. The transition region requires special treatment, and we will turn to that in the next subsection.

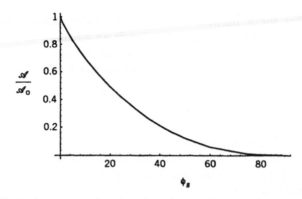

Figure 2.21. Bucket area as a function of ϕ_s relative to the area of a stationary bucket.

2.2.3 Transition Crossing

We begin this treatment by rewriting Equation 2.61 for a circular accelerator. We note that the angular frequency of the RF system is $\omega_{rf} = hv/R$, where h is the harmonic number and R is the circumference divided by 2π, and that $v\tau = 2\pi R$. Then, we obtain

$$\frac{d^2\Delta\phi}{dt^2} - \frac{1}{\eta/E_s}\frac{d}{dt}\left(\frac{\eta}{E_s}\right)\frac{d\Delta\phi}{dt} - \left(\frac{hc^2eV\cos\phi_s}{2\pi R^2}\right)\left(\frac{\eta}{E_s}\right)\Delta\phi = 0. \quad (2.77)$$

Let's suppose that we cross transition at $t = 0$ in such a way that

- $d(\eta/E_s)/dt = k$, a constant, and
- $eV|\cos\phi_s| = $ constant.

Naturally, this implies that the synchronous phase jump occurs precisely at transition. Then Equation 2.77 becomes

$$\frac{d^2\Delta\phi}{dt^2} - \frac{1}{t}\frac{d\Delta\phi}{dt} - \left(\frac{khc^2eV\cos\phi_s}{2\pi R^2}\right)t\,\Delta\phi = 0. \quad (2.78)$$

Under our two assumptions above, this equation of motion is exact. We expect its solutions to match at large t the forms of the preceeding subsection. If we cross transition holding dE_s/dt constant, our Equation 2.78 remains a good approximation.

Consider the regimes $t > 0$ and $t < 0$ separately. For $t > 0$, we have $\cos \phi_s < 0$ and so

$$\frac{d^2\Delta\phi}{dt^2} - \frac{1}{t}\frac{d\Delta\phi}{dt} + \left(\frac{khc^2eV|\cos\phi_s|}{2\pi R^2}\right)t\,\Delta\phi = 0, \qquad (2.79)$$

whereas for $t < 0$, we have $\cos \phi_s > 0$, and if we define $t_- \equiv -t$, then

$$\frac{d^2\Delta\phi}{dt_-^2} - \frac{1}{t_-}\frac{d\Delta\phi}{dt_-} + \left(\frac{khc^2eV\cos\phi_s}{2\pi R^2}\right)t_-\,\Delta\phi = 0. \qquad (2.80)$$

We see that the two equations are formally identical. Our equation is a special case of the general form

$$u'' + \frac{1 - 2a}{z}u' + \left[(qgz^{g-1})^2 + \frac{a^2 - \nu^2 g^2}{z^2}\right]u = 0, \qquad (2.81)$$

where $u = u(z)$ and $u' \equiv du/dz$, and has solutions

$$u(z) = z^a J_\nu(qz^g), \qquad (2.82)$$

$$u(z) = z^a N_\nu(qz^g), \qquad (2.83)$$

where J_ν and N_ν are Bessel and Neumann functions of order ν respectively.[1] For our special case, $a = 1$, $g = \frac{3}{2}$, $\nu = \frac{3}{2}$, and

$$q = \sqrt{\frac{4hc^2eV|\cos\phi_s|k}{9 \times 2\pi R^2}}. \qquad (2.84)$$

Under our assumptions, the solution is

$$\Delta\phi(t) = A|t|J_{2/3}(q|t|^{3/2}) + B|t|N_{2/3}(q|t|^{3/2}), \qquad (2.85)$$

where A and B are constants of integration.

---

[1] See, for example, I. S. Gradshteyn and I. M. Ryzhik, *Table of Integrals, Series, and Products*, Academic Press, New York, 1980, Equation 8.491-3.

We should check that our solution has the proper asymptotic limits as $t \to \pm\infty$. The asymptotic limits of the Bessel functions are

$$J_\nu(z) \to \sqrt{\frac{2}{\pi z}} \cos\left(z - \tfrac{1}{2}\nu\pi - \tfrac{1}{4}\pi\right), \tag{2.86}$$

$$N_\nu(z) \to \sqrt{\frac{2}{\pi z}} \sin\left(z - \tfrac{1}{2}\nu\pi - \tfrac{1}{4}\pi\right). \tag{2.87}$$

It is to be hoped that the results of the previous subsection will appear. To avoid a proliferation of absolute value signs, we neglect them with the understanding that t is a positive quantity in what follows. The amplitude becomes

$$t\sqrt{\frac{2}{\pi}} \frac{1}{\sqrt{qt^{3/2}}} \sim \frac{t^{1/4}}{q^{1/2}} \sim \left(\frac{t}{k}\right)^{1/4} \sim \left(\frac{\eta}{E_s}\right)^{1/4}, \tag{2.88}$$

which indeed agrees with Equation 2.69 when applied to a synchrotron. The phase of the oscillation should reduce to the integral of the synchrotron oscillation frequency over time. That is, we should be able to make the identification

$$qt^{3/2} = \int \Omega_s \, dt, \tag{2.89}$$

implying that

$$\Omega_s = \tfrac{3}{2} q t^{1/2}, \tag{2.90}$$

and indeed, inserting the expression for q, we find that we have Ω_s as given by Equation 2.50 when applied to a synchrotron.

Next we look at $\Delta\phi$ at transition ($t = 0$). The limiting forms of the Bessel functions when the arguments approach zero are

$$J_\nu(z) \to \frac{(z/2)^\nu}{\Gamma(\nu + 1)}, \tag{2.91}$$

$$N_\nu(z) \to -\frac{\Gamma(\nu)}{\pi(z/2)^\nu}. \tag{2.92}$$

So one of the terms goes to zero:

$$tJ_{2/3}(qt^{3/2}) \to t\frac{(q/2)^{2/3}t}{\Gamma(\tfrac{5}{3})} \to 0. \tag{2.93}$$

But the second term remains finite:

$$tN_{2/3}(qt^{3/2}) \rightarrow -t\frac{\Gamma(\frac{2}{3})}{\pi(q/2)^{2/3}t} \rightarrow -\frac{2^{2/3}\Gamma(\frac{2}{3})}{\pi q^{2/3}}, \qquad (2.94)$$

and so our solution for $\Delta\phi$ at transition is

$$\Delta\phi(0) = -\frac{2^{2/3}\Gamma(\frac{2}{3})B}{\pi q^{2/3}}. \qquad (2.95)$$

Now we must examine the corresponding behavior for ΔE, or more conveniently, $\tau\Delta E$. Since

$$\tau\Delta E = \frac{2\pi R^2}{hc^2}\frac{1}{\eta/E_s}\frac{d\Delta\phi}{dt}, \qquad (2.96)$$

we will evaluate the derivative of our solution for $\Delta\phi$. We obtain

$$\frac{d\Delta\phi}{dt} = AJ_{2/3}(qt^{3/2}) + AtJ'_{2/3}(qt^{3/2})\tfrac{3}{2}qt^{1/2} \qquad (2.97)$$

$$+ BN_{2/3}(qt^{3/2}) + BtN'_{2/3}(qt^{3/2})\tfrac{3}{2}qt^{1/2}. \qquad (2.98)$$

By using the identity

$$J'_{2/3}(z) = -J_{5/3}(z) + \frac{2}{3}\frac{1}{z}J_{2/3}(z) \qquad (2.99)$$

and the like form for the Neumann function, we find that

$$\frac{d\Delta\phi}{dt} = A\left[2J_{2/3}(qt^{3/2}) - \tfrac{3}{2}qt^{3/2}J_{5/3}(qt^{3/2})\right] \qquad (2.100)$$

$$+ B\left[2N_{2/3}(qt^{3/2}) - \tfrac{3}{2}qt^{3/2}N_{5/3}(qt^{3/2})\right]. \qquad (2.101)$$

We leave it to the reader to show that this solution approaches the correct asymptotic limit as $t \rightarrow \infty$; in particular, the amplitude of $\tau\Delta E$ varies as

$(E_s/\eta)^{1/4}$, as obtained earlier. The limiting case for $t \to 0$ entails a little more algebra. One will find that

$$\frac{d\Delta\phi}{dt} = \frac{2A}{\Gamma(\frac{5}{3})}\left(\frac{q}{2}\right)^{2/3} t - \frac{3}{2}\frac{A}{\Gamma(\frac{8}{3})}q\left(\frac{q}{2}\right)^{5/3}t^4 \tag{2.102}$$

$$- 2B\frac{\Gamma(\frac{2}{3})}{\pi(q/2)^{2/3}t} + \tfrac{3}{2}Bqt^{3/2}\frac{\Gamma(\frac{5}{3})}{\pi(q/2)^{5/3}t^{5/2}}. \tag{2.103}$$

If we recognize that $\Gamma(\frac{5}{3}) = (\frac{2}{3})\Gamma(\frac{2}{3})$, then the third and fourth terms cancel.

In Equation 2.96 the quantity η/E_s in the denominator is proportional to t. Hence, when multiplied by our expression for $d\Delta\phi/dt$, the second term will vanish for t approaching zero. The first term, however, will remain and becomes

$$\tau\,\Delta E(0) = \frac{2^{1/3}A(2\pi R^2)q^{2/3}}{hc^2\Gamma(\frac{5}{3})k}. \tag{2.104}$$

These expressions for the phase and energy deviations at transition tell us that though a particular particle may reach $\Delta\phi = 0$ at transition due to its initial conditions, the maximum values for $\Delta\phi$ and ΔE for an ensemble of particles are not zero and infinity, respectively. Consider an ensemble of particles distributed along a common ellipse in longitudinal phase space in the adiabatic limit. For this ensemble, each particle will have different initial conditions characterized by different values of A and B. But, for all particles, $A^2 + B^2$ will have the same value. The maximum value of ΔE at transition will occur for the particle which has $B = 0$. Note that A for this particular particle characterizes the amplitude of the energy oscillation for large values of t as well. Therefore, we may compare the maximum energy excursion at transition with the maximum energy excursion at large t. To do so, we must write out $\tau\,\Delta E$ for large t explicitly. We obtain

$$\tau\,\Delta E = \frac{3\pi R^2}{hc^2}\sqrt{\frac{2}{\pi}}\left[\frac{4hc^2eV|\cos\phi_s|E_s}{9\times2\pi R^2k^2|\eta|}\right]^{1/4}\left[-A\sin\theta_{2/3} + B\cos\theta_{2/3}\right], \tag{2.105}$$

where

$$\theta_{2/3} = \int\Omega_s\,dt - \frac{2}{3}\frac{\pi}{2} - \frac{\pi}{4}. \tag{2.106}$$

For completeness,

$$\Delta\phi = \sqrt{\frac{2}{\pi}}\left[\frac{9\times2\pi R^2|\eta|}{4hc^2eV|\cos\phi_s|E_sk^2}\right]^{1/4}\left[A\cos\theta_{2/3} + B\sin\theta_{2/3}\right]. \tag{2.107}$$

Comparing A for the particle of interest, we get

$$A_{max} = \frac{(\tau \Delta E)_i}{f_i} = \frac{(\tau \Delta E)_t}{f_t},$$ (2.108)

where $(\tau \Delta E)_i$ and $(\tau \Delta E)_t$ are the initial value of $\tau \Delta E$ and its value at transition, respectively, and

$$f_i = \frac{3\pi R^2}{hc^2} \sqrt{\frac{2}{\pi}} \left[\frac{4hc^2 eV |\cos\phi_s| E_s}{9 \times 2\pi R^2 k^2 \eta} \right]_i^{1/4},$$ (2.109)

$$f_t = \frac{2^{1/3}(2\pi R^2) q^{2/3}}{hc^2 \Gamma\left(\frac{5}{3}\right) k}.$$ (2.110)

If evaluated at transition, the constant k can be expressed as

$$k = \frac{2eV \sin \phi_s}{\tau_t \gamma_t^2 E_t^2}.$$ (2.111)

The meaningful quantities to compare are not necessarily the $\tau \Delta E$'s, but rather the fractional energy differences $\Delta E/E$ at the two times, because it is these fractional energy differences that relate directly to aperture demand. With this change, in combination with the above results,

$$\frac{\Delta E_t/E_t}{\Delta E_i/E_i} = \frac{1}{\Gamma\left(\frac{5}{3}\right)} \left(\frac{v_t}{v_i}\right) \left[\frac{2^{11}\pi^7}{3^{14}} \frac{hc^2}{v_t^2} \frac{\cos \phi_s}{\sin^2 \phi_s} \frac{E_i^9}{E_0^4 E_t^4 eV} |\eta_i|^3 \right]^{1/12}.$$ (2.112)

In this expression, $E_0 = mc^2$ is the rest energy of the particle, and the voltage V and synchronous phase ϕ_s are to be evaluated at transition.

Since $\Delta E/E_s$ does not become infinite, how big does it get? Let's put in some numbers corresponding to the Main Ring at Fermilab, which crosses transition. Our model with its constant time derivative of η/E_s does not correspond to the true operating cycle of that accelerator, but here we are only interested in an order of magnitude estimate. We take $h = 1113$, $\phi_s = 45°$, $\gamma_t = 18$, $eV = 1$ MeV, $E_i = 9$ GeV and obtain

$$\frac{\Delta E_t/E_t}{\Delta E_i/E_i} = 1.11.$$ (2.113)

That is, the maximum energy spread at transition is only 10% larger than the energy spread at injection under these assumptions.

We conclude with an observation about one of the difficulties in crossing transition. Far from transition, the beam is relatively forgiving of errors in the acceleration frequency, because so long as those errors are not too large, the beam will simply move to a new equilibrium radius. In the vicinity of transition, however, the beam is completely unforgiving of frequency errors. Since there is no synchrotron oscillation frequency, the wrong frequency could drive the beam into the vacuum chamber walls. Typically feedback systems are employed to match the acceleration system to the actual beam frequency.

2.3 THE NEED FOR TRANSVERSE FOCUSING

We have shown how a radiofrequency electromagnetic field can be used to accelerate charged particles in such a way that stable oscillations about the design energy will be maintained. However, if these were the only fields acting on the particles, motion in at least one direction transverse to the general direction of motion necessarily would be unstable. To see this, consider the radiofrequency (RF) wave in a linac structure as shown in Figure 2.22. Here, we plot the RF voltage as a function of time as seen at a particular longitudinal location z.

The synchronous particle arrives at z_s when the voltage is at a value $V \sin \phi_s$. Particles just behind the synchronous particle arrive late and therefore see a higher accelerating field, and we have shown that this results in stable longitudinal motion for a linac or for a synchrotron below the transition energy. If we now look at the electric field of the same wave as a function of z instead of t, the picture becomes that in Figure 2.23.

Upon transforming to the rest frame of the synchronous particle, the wave would appear stationary; the magnetic field would be zero and the z-component of the electric field would remain unchanged. In this frame the z-component of the electric field will still have a negative gradient: $\partial E_z / \partial z < 0$. But, since $\nabla \cdot \vec{E} = 0$ in this region (neglecting the fields due to the particles themselves), then there must be at least one transverse component of the

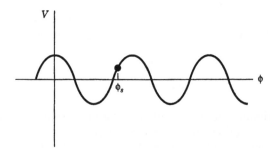

Figure 2.22. RF voltage as a function of time as seen at a particular longitudinal location z.

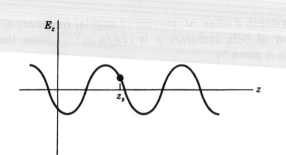

Figure 2.23. Longitudinal electric field as a function of longitudinal location z, as seen at a particular time t.

electric field which has a positive gradient. In fact, if the geometry of the accelerating station is cylindrically symmetric, then

$$\frac{\partial E_r}{\partial r} + \frac{\partial E_z}{\partial z} = 0, \tag{2.114}$$

or

$$\frac{\partial E_r}{\partial r} > 0. \tag{2.115}$$

Hence, the radial forces that the particles experience will be outward and increasing with radial position; the resulting transverse motion will be unstable. (Upon transforming back to the lab frame, one would find that the effect decreases with increasing energy, due to the magnetic forces. At highly relativistic energies, the net force would be negligible.)

Though the effect just described is most relevant in a proton linear accelerator, similar arguments will be made to show that circular accelerators could not operate without transverse focusing (even if they were not accelerating!). Other effects, such as the space charge forces between individual particles, forces on the particles due to image charges in the vacuum chamber, and so forth, will contribute to the transverse behavior of the particles. The motion in the transverse degrees of freedom and its stability will be the subject of the next chapter.

PROBLEMS

1. A betatron has accelerated electrons to an energy of 300 MeV. If the orbit radius was 1 meter, calculate the field at the orbit and the flux through the orbit at 300 MeV. Given the circumstance that magnet iron saturates at about 2 tesla, estimate the cross section of the magnet yoke. Comment on the feasibility of a 10 GeV betatron.

2. The skin depth δ of an AC current of angular frequency ω traveling on a conductor of bulk resistivity ρ is $(2\rho/\mu_0\omega)^{1/2}$. Show that the surface resistivity is given by

$$\rho_s = \left(\frac{\omega\mu_0\rho}{2}\right)^{1/2}.$$

The unit of surface resistivity is the ohm, but it is frequently stated in the engaging unit of "ohms per square." Calculate the surface resistivity for copper at $f = 400$ MHz.

3. Find the ratio R/L of a pillbox cavity for which the shunt impedance R_s of a pillbox cavity is a maximum. If the shunt impedance per unit length is defined by $r_s \equiv R_s/L$, find the ratio of R/L for which r_s is a maximum.

4. For many years, surface breakdown fields under cw conditions have been estimated with the empirical Kilpatrick criterion, established when untrapped oil diffusion vacuum pumps were used. The criterion can be written in the form

$$f = 1.64E_k^2 \exp(-8.5/E_k).$$

Here, the frequency f is in MHz, and the electric field E_k, is in MV/m. Nowadays, it is reasonable to design for fields above the Kilpatrick limit by close to a factor of two. Using a maximum surface field of $1.7E_k$, find the energy gain possible for the pillbox cavity treated in the text. If a synchrotron is to produce acceleration at the rate of 3 MeV per turn, how many such cavities would be required, and what would be their total power dissipation?

5. From the first integral of the equation of motion, show that the range of stable phases is from $\phi_1 = \pi - \phi_s$ to the solution ϕ_2 of the equation

$$\cos \phi_2 + \phi_2 \sin \phi_s = -\cos \phi_s + (\pi - \phi_s)\sin \phi_s.$$

6. Using a graphics terminal, produce turn by turn plots in $\Delta E/E_s$-ϕ coordinates from the original difference equations, and compare with the mappings shown from the solutions of the differential equations.

7. A set of acceleration parameters for a large synchrotron might run as follows. A ring with a circumference of 87 km accelerates protons from 2 to 20 TeV in 1500 seconds. The acceleration system has a "voltage" amplitude of 15 MV and operates at 360 MHz. Calculate the synchronous phase and the frequency of small amplitude phase oscillations, assuming $\gamma \gg \gamma_t$. Use $\gamma_t = 105$.

8. Derive the expression for the area of a stationary bucket. Evaluate the area for a ring with a circumference of 2π km operating with $h = 1113$, for $E_s = 150$ GeV and $V = 1$ MV. Here, use $\gamma_t = 18$.

9. Derive the expression for the longitudinal emittance of a beam with a maximum phase oscillation amplitude ϕ_m within a stationary bucket. Use this expression to estimate the longitudinal emittance of a bunch with $\phi_m = 0.5$ in the synchrotron of the preceding problem.

10. Consider a beam of particles with maximum phase deviation given by ϕ_m and maximum momentum deviation given by $(\Delta p/p_s)_m$. Show how $(\Delta p/p_s)_m$ varies with energy when (a) $\gamma \ll \gamma_t$, and when (b) $\gamma \gg \gamma_t$.

11. For a stationary bucket, show that the ratio of the synchrotron oscillation frequency for a particle with phase amplitude ϕ_m to that of a particle with small phase amplitude is

$$\frac{\nu_s(\phi_m)}{\nu_s(0)} = \frac{\pi}{2} \frac{1}{K\left(\sin^2 \frac{\phi_m}{2}\right)}.$$

where K is the complete elliptic integral of the first kind:

$$K(x) \equiv \int_0^{\pi/2} (1 - x \sin^2 \theta)^{-1/2} \, d\theta.$$

12. This is a digression, but it's interesting to note the similarity between the difference equations for synchrotron oscillations and a mapping that is frequently used in mathematical studies in nonlinear dynamics. By a suitable change of variables, transform the difference equations for a stationary bucket into the form

$$\theta_{n+1} = \theta_n + r_{n+1},$$

$$r_{n+1} = r_n - \frac{k}{2\pi} \sin 2\pi\theta_n,$$

for $0 \le r$ and $\theta \le 1$, and with the understanding that as θ and r are incremented only the fractional parts are retained. With the aid of a graphics terminal, exhibit the motion in θ, r for $k = 0.1$. You should see behavior familiar from the synchrotron motion plots. Now raise k to 0.9 and start a particle at, for example, $r = 0.27$ and $\theta = 0$. The plot quickly develops an unpredictable and chaotic character. Fortunately, the equivalent acceleration system parameters are far from those actually used in synchrotrons.

13. Calculate the maximum relative energy deviation $\Delta E / E_s$ which can be contained in a stationary bucket at injection into the SSC collider. Use an injection energy of 2 TeV, and RF voltage of 8 MV, harmonic number of 10^5, and a transition gamma of 105.

14. In particle–antiparticle synchroton colliders, the two beams usually share a single magnet ring. One may wish to manipulate the energy of one beam independently of the other, although both beams pass through the same radiofrequency accelerating system. Propose a method of achieving this goal. Assume that you have freedom to install as many RF stations as you wish at locations of your choice.

CHAPTER 3

Transverse Linear Motion

In the previous chapter we concentrated on acceleration and energy stability and found a stability principle that causes particles initially near each other in energy to remain so in the presence of oscillatory accelerating fields. We referred to the associated energy oscillations as motion in the longitudinal degree of freedom. Reasonably enough, the other two degrees of freedom are termed transverse, as indicated in Figure 3.1.

One of our tasks in this chapter is to investigate transverse stability. Do particles of the same energy, but with slightly different transverse coordinates, either in position or direction, remain near each other in the course of their motion in the accelerator? We will find a criterion for such stability and discuss the solution of the associated equations of motion. We will show that, in general, stability consists in bounded oscillatory motion about the design trajectory. This motion is termed a *betatron oscillation* for historical reasons. We will see that the transverse oscillation frequencies are much greater than the typical frequency of phase oscillations, thus allowing us to treat the longitudinal degree of freedom independently. We will thus have identified stability principles for all three degrees of freedom which are passive in the sense that, for our ideal accelerator, they do not rely on feedback mechanisms.

In a linear accelerator, particles of different momenta follow the same ideal trajectory. In a circular accelerator this is not the case, and we will exhibit the closed orbits of particles differing in momenta from that of the ideal particle. Thus, for this variety of accelerator, there is a transverse attribute to phase oscillations.

We can choose our ideal accelerator such that the transverse restoring forces are linear in the transverse coordinates. For high energy accelerators, the restoring forces and the bending forces in circular accelerators are

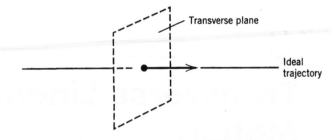

Figure 3.1. Characterization of transverse and longitudinal degrees-of-freedom of particle motion.

produced by magnetic rather than electric fields. For a nonrelativistic parti-cle, in the low speed limit, electrostatic fields are more effective. But for a particle traveling near the speed of light, a magnetic field of 1 T and an electric field of 300 MV/m would each provide the same transverse deflect-ing force. The former field strength is typical of magnetic devices, while the latter is outside our present reach.

With our ideal linear restoring forces, the transverse degrees of freedom are also independent of one another; however, the two transverse frequen-cies are comparable and hence certain imperfections in the restoring fields can couple the motion. This topic of transverse coupled motion is left to a separate chapter. We conclude this chapter with an introduction to perturba-tions to the ideal linear magnetic fields which cause steering and focusing errors.

3.1 STABILITY OF TRANSVERSE OSCILLATIONS

The most basic magnetic field configuration—a uniform field— produces a form of focusing. Consider a particle traveling along a circular orbit in such a

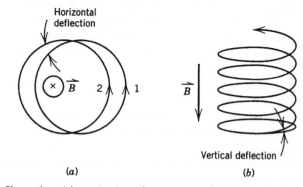

Figure 3.2. Charged particle motion in uniform magnetic field when perturbed by a deflection (a) perpendicular to the field lines, and (b) parallel to the field lines.

field. Suppose the particle receives an angular deflection in the plane perpendicular to the magnetic field. The resulting orbit will be just another circle of the same radius as the first but with a different center, as illustrated in Figure 3.2. One can say that the second orbit is performing a stable oscillation about the first. Unfortunately, if the deflection has a component along the magnetic field lines, the particle will subsequently spiral away without limit—there is no focusing in this degree of freedom.

3.1.1 Weak Focusing

The situation of Figure 3.2 can be rectified by designing the source of the magnetic field so that the field lines bend outward as shown in Figure 3.3. Particles above the midplane will experience a force downward; those below will be forced upward. However, along the horizontal plane the vertical component of the magnetic field decreases with increasing radius, since the field lines get farther apart. Thus vertical focusing is achieved at the expense of radial focusing, and so there is a limit to the effectiveness of the focusing that can be achieved in both transverse degrees of freedom simultaneously.

Suppose the vertical component of the field along the midplane is given by

$$B_y = \frac{B_0}{r^n}. \tag{3.1}$$

Then, if $n = 0$, we would have a uniform field and no vertical focusing: just the situation discussed earlier. If $n > 1$, the field could not provide the necessary centripetal force to keep the particles moving in a circular path of constant radius. Hence, for stability, the field index n is constrained to lie between the values of 0 and 1.

This form of focusing is called *weak focusing*. It has the disadvantage that as the design energy, and hence the circumference of the orbit, is increased, so also does the required aperture increase for a given angular deflection. Because the focusing is weak, the radial oscillations are essentially those depicted in Figure 3.2(a); the maximum radial displacement of a deflected particle is directly proportional to the radius of the machine. The scale of the magnetic components of a synchrotron, for example, would become unreasonably large and costly. This circumstance led to the invention of alternating gradient focusing (also known as *strong focusing*) in 1952. This method is the

Figure 3.3. Cross section of weak focusing circular accelerator.

one most commonly used in accelerators today and we will consider its properties for the remainder of this chapter.

3.1.2 Strong Focusing

One would like the restoring force on a particle displaced from the design trajectory to be as strong as possible. The weak focusing scheme described above is limited. In the absence of current density, field gradients that provide restoring forces in both transverse degrees of freedom simultaneously are not possible. The condition $\vec{\nabla} \times \vec{B} = 0$ leads to

$$\frac{\partial B_y}{\partial x} = \frac{\partial B_x}{\partial y}, \tag{3.2}$$

where x and y are the two transverse coordinates. For small displacements, x and y, from the design trajectory, the field may be written as

$$\vec{B} = B_x \hat{x} + B_y \hat{y} \tag{3.3}$$

$$= \left(B_x(0,0) + \frac{\partial B_x}{\partial y} y + \frac{\partial B_x}{\partial x} x \right) \hat{x} + \left(B_y(0,0) + \frac{\partial B_y}{\partial x} x + \frac{\partial B_y}{\partial y} y \right) \hat{y}, \tag{3.4}$$

where $\hat{x}$ and $\hat{y}$ are unit vectors in the x and y directions, respectively. The last term in each component produces a force at right angles to the displacement and hence cannot represent a restoring force. The remaining coefficients of x and y are equal, according to the curl condition, Equation 3.2. Then the Lorentz force is focusing in one coordinate and defocusing in the other. The standard magnet that produces this focusing character is the quadrupole.

The focal length of a *thin lens* quadrupole magnet can be obtained easily. We imagine a charged particle moving through the quadrupole at a distance x from the magnet's axis of symmetry. The thin lens approximation implies that the length of the magnet, l, is short enough that the displacement x is unaltered as the particle passes through the magnet and hence the magnetic field experienced by the particle, $B_y = (\partial B_y / \partial x) x$, is constant along the particle trajectory. In this *paraxial* approximation, the angle is equal to the slope of the particle's trajectory, $x' \equiv dx/ds$ (where s is distance measured along the ideal trajectory). As depicted in Figure 3.4, the slope of the particle's transverse trajectory thus will be altered by an amount

$$\Delta x' = -\frac{l}{\rho} = -l \cdot \left(\frac{eB_y}{p} \right) = -\left(\frac{eB'l}{p} \right) x, \tag{3.5}$$

where ρ is the radius of curvature of the trajectory through the magnetic field

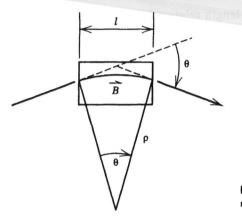

Figure 3.4. Deflection of particle by thin magnetic element.

and $B' \equiv \partial B_y/\partial x$ is the *gradient* of the quadrupole magnet, and where use has been made of Equation 2.1.

Since a ray parallel to the optic axis will be bent toward the focal point of the lens, as depicted in Figure 3.5, the change in slope is simply $\Delta x' = -x/f$, where f is the focal length of the quadrupole lens. The focal length is thus given by

$$\frac{1}{f} = \frac{eB'l}{p}. \tag{3.6}$$

The ratio of momentum to charge, p/e, is often called the *magnetic rigidity* and written $(B\rho)$; we will follow this latter convention in much of the text. Remember that $(B\rho)$ is just a single symbol. In the MKS system of units, $(B\rho)$ can be calculated from

$$(B\rho) = \frac{10}{2.9979} p_{(\text{GeV}/c)} \qquad \text{tesla-meters.} \tag{3.7}$$

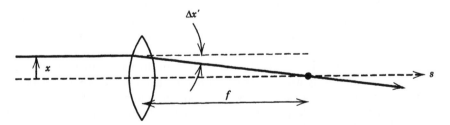

Figure 3.5. Ray initially parallel to the optical axis is bent by a convex lens causing it to pass through its focal point a distance f away.

We can therefore express our focal length relationship by

$$\frac{1}{f} = \frac{B'l}{(B\rho)}. \tag{3.8}$$

As mentioned earlier, quadrupole lenses focus in one plane and defocus in the other. Obviously, an accelerator cannot be made up of magnets that focus only in one plane. But we can recall from geometrical optics that a combination of equal strength convex and concave lenses will produce a net focusing. To see this, we recast Equation 3.5 in matrix form:

$$\begin{pmatrix} x \\ x' \end{pmatrix}_{out} = \begin{pmatrix} 1 & 0 \\ -\dfrac{1}{f} & 1 \end{pmatrix} \begin{pmatrix} x \\ x' \end{pmatrix}_{in}. \tag{3.9}$$

For a concave lens, the focal length is of opposite sign. In this language, the progress of a ray through the interlens space of length L is given by

$$\begin{pmatrix} x \\ x' \end{pmatrix}_{out} = \begin{pmatrix} 1 & L \\ 0 & 1 \end{pmatrix} \begin{pmatrix} x \\ x' \end{pmatrix}_{in}. \tag{3.10}$$

Therefore, the matrix corresponding to transport of the ray through first a concave lens, then a drift, and then a convex lens may be written as

$$\begin{pmatrix} 1 & 0 \\ -\dfrac{1}{f} & 1 \end{pmatrix} \begin{pmatrix} 1 & L \\ 0 & 1 \end{pmatrix} \begin{pmatrix} 1 & 0 \\ \dfrac{1}{f} & 1 \end{pmatrix} = \begin{pmatrix} 1 + \dfrac{L}{f} & L \\ -\dfrac{L}{f^2} & 1 - \dfrac{L}{f} \end{pmatrix}. \tag{3.11}$$

At least in the case where L is small compared to f, it is clear that there is net focusing; in this approximation the resulting matrix is that of a thin lens of net focal length $f^2/L > 0$. If the two lenses were interchanged, the net result would still be focusing. Hence a system of alternating gradient thin quadrupole magnets could, in principle, focus in both degrees of freedom at once. In fact, the focal length need not be large compared with the lens spacing for this to occur, which is the point of one of the problems at the end of the chapter.

The above discussion suggests that one can focus in two degrees of freedom simultaneously using a system of magnetic elements whose gradients alternate in sign. The discussion was based on tracing the trajectories of particles through a single pass of a system of two lenses of alternating focal lengths. In modern accelerators, where particles must be transported through great distances, the stability of particle motion through repetitive encounters with such structures must be studied. Below we develop a criterion for stable motion through an infinite number of passages through a focusing structure and apply the result to the thin lens alternating gradient system. The type of focusing achieved using alternating gradients is called *strong focusing*. As we shall see, strong focusing leads to beam sizes which are dependent upon the spacing of the lenses and their strengths, and independent of the scale of the accelerator.

3.1.3 Stability Criterion

In a synchrotron or long beam transport composed of alternately focusing and defocusing lenses, it is not obvious at the outset what relationships between lens strengths and spacing lead to stable oscillations as opposed to oscillations that grow in amplitude with time. The matrix language introduced in the preceding section can be used to establish a condition which distinguishes between these alternatives.

The detailed description of the way in which magnets and intervening spaces are placed to form the accelerator is conventionally called the *lattice*. After having developed the matrices appropriate to the elements of the accelerator, the motion of a particle can be followed through the lattice. If a particle traverses a series of elements having matrices $M_1, M_2, \ldots, M_n$, then the input and output conditions through these elements are related by the matrix

$$M = M_n \cdots M_2 M_1.$$

If the sequence of elementary matrices above is encountered repetitively, as is the case, for instance, if they represent the components all the way around a circular accelerator, then we can use this matrix to enquire into the stability of transverse oscillations.

For an oscillation to be stable, the quantity

$$M^n \begin{pmatrix} x \\ x' \end{pmatrix}_{in}, \tag{3.12}$$

where M is the matrix for one turn or repetition period, must remain finite for arbitrarily large n. Let V_1, V_2 be the two eigenvectors of M, corresponding to the eigenvalues λ_1, λ_2. Any initial condition can be expressed in terms of V_1, V_2:

$$\begin{pmatrix} x \\ x' \end{pmatrix}_{in} = A V_1 + B V_2, \tag{3.13}$$

where A and B are constants. Propagation for n periods is then represented by

$$M^n \begin{pmatrix} x \\ x' \end{pmatrix}_{in} = A \lambda_1^n V_1 + B \lambda_2^n V_2, \tag{3.14}$$

and so the requirement for stability is equivalent to the requirement that λ_1^n and λ_2^n not grow with n.

But note that M is the product of matrices each of which has determinant equal to unity, so M itself is unimodular. The eigenvalues of M are thus reciprocals of each other:

$$\lambda_2 = 1/\lambda_1, \tag{3.15}$$

and we can in general write

$$\lambda_1 = e^{i\mu},$$
$$\lambda_2 = e^{-i\mu},$$

where μ is a complex number. For stability, we see that μ must be real.

Now, solve the eigenvalue equation for M. Setting

$$M = \begin{pmatrix} a & b \\ c & d \end{pmatrix}$$

the eigenvalue equation

$$\det(M - \lambda I) = 0 \tag{3.16}$$

becomes

$$(ad - bc) - (a + d)\lambda + \lambda^2 = 0. \tag{3.17}$$

Noting that $ad - bc = \det M = 1$ and rearranging gives

$$\lambda^{-1} + \lambda = a + d \equiv \operatorname{Tr} M, \tag{3.18}$$

where $\operatorname{Tr} M$ stands for the trace of M. Finally, expressing λ in terms of μ gives

$$e^{i\mu} + e^{-i\mu} = 2 \cos \mu = \operatorname{Tr} M, \tag{3.19}$$

and the stability condition is just

$$-1 \le \tfrac{1}{2} \operatorname{Tr} M \le 1. \tag{3.20}$$

Note that the stability condition is independent of the starting point, since the trace of a product of matrices is invariant under cyclic permutation of the matrices.

The significance of the angle μ will appear in the next section, where it will be identified as the phase advance of the transverse oscillation through the interval contained in M.

As an example, consider a lattice which consists only of equally spaced focusing and defocusing lenses, each of which we will assume to be thin. (This is referred to as a FODO lattice.) If the order is first the focusing lens, then a drift of length L, third a defocusing lens, and finally another drift of length L, the matrix is

$$M = \begin{pmatrix} 1 - \dfrac{L}{f} - \left(\dfrac{L}{f}\right)^2 & 2L + \dfrac{L^2}{f} \\ -\dfrac{L}{f^2} & 1 + \dfrac{L}{f} \end{pmatrix}. \tag{3.21}$$

Application of the stability condition gives

$$-1 \le 1 - \frac{1}{2}\left(\frac{L}{f}\right)^2 \le 1,$$

or, simplifying,

$$\left|\frac{L}{2f}\right| \le 1. \tag{3.22}$$

So we obtain the remarkably simple result that the motion is stable provided the focal length is greater than half the lens spacing.

Now we are in a position to suggest one of the significant advantages of strong focusing. In Figure 3.6, we sketch an oscillation of a particle traversing a sequence of focusing and defocusing lenses for the case that the focal length is half the lens spacing. We see that the wavelength of an oscillation is just four times the lens spacing and therefore is unrelated to the size of the accelerator. The alternating gradient principle enables us to decouple the transverse aperture requirement from the size (hence, energy) of the accelerator. Even though for simplicity of the illustration we have made this

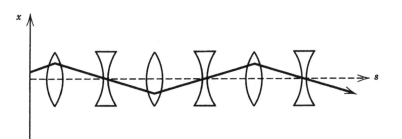

Figure 3.6. Example of a particle oscillation through a system of lenses where $f = L/2$. The maximum displacement is independent of the size of the accelerator.

argument at the limit of stability, it of course remains true for focal lengths within the stability bounds.

3.2 EQUATION OF MOTION

We are dangerously close to being able to write down the differential equations of motion for transverse oscillations. Consider a particle passing through a magnetic field with gradient $B' = \partial B_y/\partial x$ over a distance Δs. Recalling Equation 3.5, we see that the slope of a particle's trajectory, $x' = dx/ds$, changes by an amount $\Delta x' = -[B'\Delta s/(B\rho)]x$. Thus,

$$\frac{\Delta x'}{\Delta s} = -\frac{B'(s)}{(B\rho)}x. \qquad (3.23)$$

Taking the limit as $\Delta s \to 0$, we obtain the second order differential equation

$$x'' + \frac{B'(s)}{(B\rho)}x = 0. \qquad (3.24)$$

If there were a nonzero magnetic field on the design trajectory, as in a bending magnet, then Equation 3.24 would represent the difference between the slope changes of the particle in question and that of the ideal particle. While we have obtained the essentials of the equation of motion in just two steps, we have an obligation to the reader to provide a more rigorous derivation, which we now proceed to do.

Let us limit ourselves to the situations in which the design trajectory is a straight line or a single planar closed curve. By implication, we are developing the equation of motion for betatron oscillations in a linac or synchrotron.

Suppose that the geometry is as sketched in Figure 3.7. Locally, the design trajectory (reference orbit) has curvature ρ. The path length along this curve is s. Ultimately, s will be the independent variable. At any point along the reference orbit, we can define three unit vectors: $\hat{s}, \hat{x}, \hat{y}$. The position of a particle can then be expressed as a vector $\vec{R}$ in the form

$$\vec{R} = r\hat{x} + y\hat{y}, \qquad (3.25)$$

where $r \equiv \rho + x$. We are interested in the behavior of the deviations x and y from the reference orbit.

The equation of motion is

$$\frac{d\vec{p}}{dt} = e\vec{v} \times \vec{B}, \qquad (3.26)$$

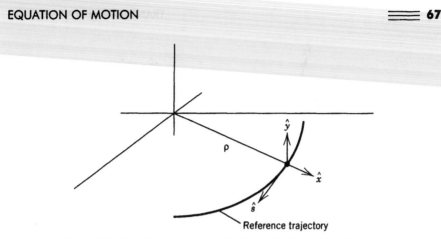

Figure 3.7. Coordinate system for development of equation of motion.

and we assume that there are radial and vertical components of $\vec{B}$—we will ignore the possible $\hat{s}$-component of $\vec{B}$ for now. So the cross product on the right hand side becomes

$$\vec{v} \times \vec{B} = \begin{vmatrix} \hat{x} & \hat{y} & \hat{s} \\ v_x & v_y & v_s \\ B_x & B_y & 0 \end{vmatrix} = -v_s B_y \hat{x} + v_s B_x \hat{y} + (v_x B_y - v_y B_x)\hat{s}. \quad (3.27)$$

If we ignore the radiation created by an accelerating charge for now, then the energy and hence the Lorentz factor γ do not change in a static magnetic field. (Synchrotron radiation due to accelerating charges will be discussed in a later chapter.) The left hand side thus becomes

$$\frac{d\vec{p}}{dt} = \frac{d}{dt}\gamma m \dot{\vec{R}} = \gamma m \ddot{\vec{R}}, \quad (3.28)$$

and so

$$\ddot{\vec{R}} = \frac{e\vec{v} \times \vec{B}}{\gamma m}. \quad (3.29)$$

Now we must evaluate $\ddot{\vec{R}}$ in these coordinates:

$$\vec{R} = r\hat{x} + y\hat{y}, \quad (3.30)$$

$$\dot{\vec{R}} = \dot{r}\hat{x} + r\dot{\hat{x}} + \dot{y}\hat{y}. \quad (3.31)$$

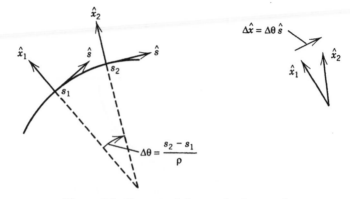

Figure 3.8. Time rate of change of unit vector $\hat{x}$.

We have to include the $\dot{\hat{x}}$-term in the above because if there is motion in the s-direction, the unit vector $\hat{x}$ will have a derivative. From Figure 3.8 we see that

$$\dot{\hat{x}} = \dot{\theta}\hat{s}, \tag{3.32}$$

where $\dot{\theta} \equiv v_s/r$. Therefore,

$$\dot{\vec{R}} = \dot{r}\hat{x} + r\dot{\theta}\hat{s} + \dot{y}\hat{y}, \tag{3.33}$$

and differentiating again,

$$\ddot{\vec{R}} = \ddot{r}\hat{x} + (2\dot{r}\dot{\theta} + r\ddot{\theta})\hat{s} + r\dot{\theta}\dot{\hat{s}} + \ddot{y}\hat{y}. \tag{3.34}$$

The new quantity in the above is $\dot{\hat{s}}$. By the same argument as used to obtain $\dot{\hat{x}}$, we have

$$\dot{\hat{s}} = -\dot{\theta}\hat{x}, \tag{3.35}$$

and so

$$\ddot{\vec{R}} = (\ddot{r} - r\dot{\theta}^2)\hat{x} + (2\dot{r}\dot{\theta} + r\ddot{\theta})\hat{s} + \ddot{y}\hat{y}. \tag{3.36}$$

Thus, in the $\hat{x}$-direction the equation of motion is

$$\ddot{r} - r\dot{\theta}^2 = -\frac{ev_s B_y}{\gamma m} = -\frac{ev_s^2 B_y}{\gamma m v_s}. \tag{3.37}$$

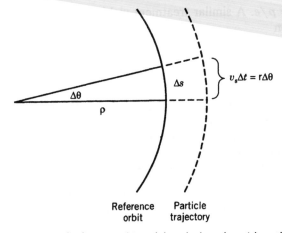

Reference   Particle
orbit   trajectory

Figure 3.9. Comparison of reference orbit path length ds and particle path length $v_s \, dt$.

Since $v_x \ll v_s$ and $v_y \ll v_s$, to a very good approximation the total momentum p of the particle is $\gamma m v_s$. So

$$\ddot{r} - r\dot{\theta}^2 = -\frac{e v_s^2 B_y}{p}. \tag{3.38}$$

Now, change to s as the independent variable. Then

$$\frac{d}{dt} = \frac{ds}{dt}\frac{d}{ds}, \tag{3.39}$$

and from Figure 3.9 we see that

$$ds = \rho \, d\theta = v_s \, dt \frac{\rho}{r}. \tag{3.40}$$

Hence, assuming $d^2s/dt^2 = 0$,

$$\frac{d^2}{dt^2} = \left(\frac{ds}{dt}\right)^2 \frac{d^2}{ds^2} = \left(v_s \frac{\rho}{r}\right)^2 \frac{d^2}{ds^2}. \tag{3.41}$$

Replacing r with $\rho + x$, the equation of motion becomes

$$\frac{d^2 x}{ds^2} - \frac{\rho + x}{\rho^2} = -\frac{B_y}{(B\rho)}\left(1 + \frac{x}{\rho}\right)^2, \tag{3.42}$$

where $(B\rho) = p/e$. A similar treatment yields for the equation of motion in the y-direction

$$\frac{d^2y}{ds^2} = \frac{B_x}{(B\rho)}\left(1 + \frac{x}{\rho}\right)^2. \tag{3.43}$$

In general, these equations will be nonlinear. But let us restrict ourselves to fields that are linear functions of x and y, and keep only the lowest order terms in x and y. Later, we will treat the higher order terms as perturbations of the basic linear motion that we consider here.

We may use the field expansion introduced in Section 3.1.1, namely

$$B_x = B_x(0,0) + \frac{\partial B_x}{\partial y}y + \frac{\partial B_x}{\partial x}x, \tag{3.44}$$

$$B_y = B_y(0,0) + \frac{\partial B_y}{\partial x}x + \frac{\partial B_y}{\partial y}y. \tag{3.45}$$

Since we are considering a planar accelerator, $B_x(0,0) = 0$. We also do not wish the motion to be coupled in the basic design, and hence the coefficients $\partial B_y/\partial y$ and $\partial B_x/\partial x$ are assumed to be zero. Finally, the equations of motion become

$$\frac{d^2x}{ds^2} + \left[\frac{1}{\rho^2} + \frac{1}{(B\rho)}\frac{\partial B_y(s)}{\partial x}\right]x = 0, \tag{3.46}$$

$$\frac{d^2y}{ds^2} - \frac{1}{(B\rho)}\frac{\partial B_y(s)}{\partial x}y = 0, \tag{3.47}$$

where use of the curl condition has been made to eliminate B_x.

These equations resemble Equation 3.24 closely; the equation in x differs only in the addition of a "centripetal" term originating in our choice of curvilinear coordinates. [This term is the origin of the focusing in x as shown in Figure 3.2(a).] For large accelerators the centripetal term is usually small in comparison with the gradient term. The equations of motion above are both of the form

$$x'' + K(s)x = 0 \tag{3.48}$$

and so differ from a simple harmonic oscillator only in that the "spring constant" K is a function of position s.

We will discuss two methods of solution. First we note that within a single magnetic component of the accelerator, K is normally a constant by design, as depicted in Figure 3.10. So within each component, we can use harmonic oscillator solutions, piecing them together at the interfaces. Second, we will examine a closed form solution using the properties of Hill's equation.

Figure 3.10. The spring constant K varies with position, but is normally constant within individual components of the accelerator.

3.2.1 Piecewise Method of Solution

In the same spirit as the geometrical optics argument of Section 3.1.2, we may describe the motion of a particle through an element of the accelerator by a 2×2 matrix. There are only three cases to consider: K vanishes, K is positive, and K is negative. The matrix for the first is the same as that for a drift space L between lenses in our earlier argument:

$$\begin{pmatrix} x \\ x' \end{pmatrix}_{\text{out}} = \begin{pmatrix} 1 & L \\ 0 & 1 \end{pmatrix}\begin{pmatrix} x \\ x' \end{pmatrix}_{\text{in}}. \tag{3.49}$$

In the y (vertical) plane, this corresponds either to a drift space between magnets or to propagation through a magnet with constant B_y. In the x (horizontal) plane, this corresponds either to a drift space between magnets or to a situation in which the centripetal term $(1/\rho^2)$ is exactly balanced by the field gradient. The latter is an unusual circumstance. Frequently, for other than exact calculation, the radius of curvature of a high energy accelerator is so large that the centripetal term may be neglected.

For $K > 0$ over a distance l, the equation of motion is just that of a simple harmonic oscillator, so in matrix form the solution is

$$\begin{pmatrix} x \\ x' \end{pmatrix}_{\text{out}} = \begin{pmatrix} \cos(\sqrt{K}l) & \dfrac{1}{\sqrt{K}}\sin(\sqrt{K}l) \\ -\sqrt{K}\sin(\sqrt{K}l) & \cos(\sqrt{K}l) \end{pmatrix}\begin{pmatrix} x \\ x' \end{pmatrix}_{\text{in}}, \tag{3.50}$$

while for $K < 0$, the corresponding result is

$$\begin{pmatrix} x \\ x' \end{pmatrix}_{\text{out}} = \begin{pmatrix} \cosh(\sqrt{|K|}l) & \dfrac{1}{\sqrt{|K|}}\sinh(\sqrt{|K|}l) \\ \sqrt{|K|}\sinh(\sqrt{|K|}l) & \cosh(\sqrt{|K|}l) \end{pmatrix}\begin{pmatrix} x \\ x' \end{pmatrix}_{\text{in}}. \tag{3.51}$$

Note that the thin lens limit emerges from the last two forms if one keeps Kl finite as $l \to 0$. In the limit, Kl tends to the reciprocal of the focal length f, as can be seen by comparison with the matrix for a thin lens in Equation 3.9.

Using the matrices above, the motion of a particle can be followed through an arrangement of accelerator elements. If a particle traverses a series of elements having matrices $M_1, M_2, \ldots, M_n$, then as stated earlier the input and output conditions through these elements are related by the matrix

$$M = M_n \cdots M_2 M_1.$$

3.2.2 Closed Form Solution

The second method of solution is based on the observation that our equation of motion is a form of Hill's equation—a differential equation studied extensively in the nineteenth century—and that general solutions can be written for it that closely resemble simple harmonic oscillations.

The equation of motion

$$x'' + K(s)x = 0 \tag{3.52}$$

has the property that though the "spring constant" K is a function of the independent variable s, for an important class of accelerators K is periodic. That is, there is a distance C such that

$$K(s + C) = K(s). \tag{3.53}$$

The repeat distance of the hardware, C, may be as large as the circumference of a synchrotron or it may be less; in any event, we will take K to be a periodic function of position. The result of nineteenth century mathematics that we will use is that the general solution of the equation of motion can be expressed in the form

$$x = Aw(s)\cos[\psi(s) + \delta], \tag{3.54}$$

where A and δ are the two constants of integration reflecting the initial conditions, and $w(s)$ can be required to be a periodic function with periodicity C. Note the similarity to the harmonic oscillator solution. For K everywhere a positive constant, we would immediately write

$$x = A \cos[\psi(s) + \delta] \tag{3.55}$$

with $\psi = \sqrt{K}s$, and A, δ the constants of integration. When K becomes a periodic function of position, the solution will differ from that for the simple harmonic oscillator by a factor representing a spatially varying amplitude and a phase which no longer develops linearly with s.

Now we must find how $w(s)$ and $\psi(s)$ are to be determined. Substitution of the general solution into the differential equation gives

$$x'' + Kx = A(2w'\psi' + w\psi'')\sin(\psi + \delta)$$

$$+A(w'' - w\psi'^2 + Kw)\cos(\psi + \delta) = 0. \qquad (3.56)$$

Since we want the functions w and ψ to be independent of δ (which depends upon the particular motion), we will require that the coefficients of the sine and cosine terms individually vanish. Multiplying the sine term by w we have

$$2ww'\psi' + w^2\psi'' = (w^2\psi')' = 0, \qquad (3.57)$$

or

$$\psi' = \frac{k}{w(s)^2}, \qquad (3.58)$$

where k is an arbitrary constant of integration.

Using this relationship between ψ and w, the coefficient of the cosine term yields the differential equation that w must satisfy:

$$w^3(w'' + Kw) = k^2. \qquad (3.59)$$

Strictly speaking, $w(s)$ need not be periodic; it only has to be a solution of Equation 3.59. But if the motion we are trying to describe is that of a particle traveling through a periodic section of the accelerator, for instance through thousands of revolutions about a circular accelerator, it is much more useful to choose the unique periodic solution for $w(s)$. Hence we will restrict our attention for now to solutions of this equation with periodicity C.

In the previous section, the matrix propagating a transverse oscillation (betatron oscillation) from one place to another in a lattice was found as a product of matrices representing basic components of the accelerator. We can also express the same matrix in terms of the parameters introduced in this section. If we rewrite Equation 3.55 as

$$x = w(s)(A_1 \cos \psi + A_2 \sin \psi) \qquad (3.60)$$

and

$$x' = \left(A_1 w' + \frac{A_2 k}{w}\right)\cos \psi + \left(A_2 w' - \frac{A_1 k}{w}\right)\sin \psi, \qquad (3.61)$$

then for the initial conditions x_0, x_0', at $s = s_0$, the constants A_1 and A_2 are

$$A_1 = \frac{x_0}{w}, \tag{3.62}$$

$$A_2 = \frac{x_0'w - x_0 w'}{k}. \tag{3.63}$$

Now by requiring that the function w be periodic over the distance C, we may write down the matrix for propagation from s_0 to $s_0 + C$. The resulting matrix equation describing the motion is

$$\begin{pmatrix} x \\ x' \end{pmatrix}_{s_0+C} = \begin{pmatrix} \cos \Delta\psi_C - \dfrac{ww'}{k}\sin \Delta\psi_C & \dfrac{w^2}{k}\sin \Delta\psi_C \\[2ex] -\dfrac{1 + (ww'/k)^2}{w^2/k}\sin \Delta\psi_C & \cos \Delta\psi_C + \dfrac{ww'}{k}\sin \Delta\psi_C \end{pmatrix} \begin{pmatrix} x \\ x' \end{pmatrix}_{s_0}. \tag{3.64}$$

The phase of the particle's oscillation advances through the repeat period by an amount

$$\psi(s_0 \rightarrow s_0 + C) \equiv \Delta\psi_C = \int_{s_0}^{s_0+C} \frac{k\, ds}{w^2(s)}. \tag{3.65}$$

Because $w(s)$ is periodic, this integral is independent of the choice of s_0.

3.2.3 Courant-Snyder Parameters

Inspection of the matrix in Equation 3.64 reveals that the function $w^2(s)$ and its derivative both scale with the arbitrary constant k. Since the motion of the particle, and in particular the advance of the phase of its motion, is what's observed, choosing a different value of k must simply lead to a different value for the function $w^2(s)$, scaled by a factor of k. Since $w^2(s)$ and its derivative are the more fundamental quantities of the problem, it is customary to define new variables

$$\beta(s) \equiv \frac{w^2(s)}{k}, \tag{3.66}$$

$$\alpha(s) \equiv -\frac{1}{2}\frac{d\beta(s)}{ds} = -\frac{1}{2}\frac{d}{ds}\left(\frac{w^2(s)}{k}\right), \tag{3.67}$$

$$\gamma \equiv \frac{1 + \alpha^2}{\beta}, \tag{3.68}$$

and rewrite the equation for one passage through the repeat period as

$$\begin{pmatrix} x \\ x' \end{pmatrix}_{s_0+C} = \begin{pmatrix} \cos \Delta\psi_C + \alpha \sin \Delta\psi_C & \beta \sin \Delta\psi_C \\ -\gamma \sin \Delta\psi_C & \cos \Delta\psi_C - \alpha \sin \Delta\psi_C \end{pmatrix} \begin{pmatrix} x \\ x' \end{pmatrix}_{s_0}. \quad (3.69)$$

Here, the phase advance is

$$\Delta\psi_C = \int_{s_0}^{s_0+C} \frac{ds}{\beta(s)}, \quad (3.70)$$

and $\beta(s)$ may be interpreted as the local wavelength of the oscillation divided by 2π. The quantities α, β, and γ are usually referred to as Courant-Snyder parameters collectively; from now on, the function β will be referred to as the amplitude function.

So the general solution to the equation of motion can be written as

$$x(s) = A\sqrt{\beta(s)} \cos[\psi(s) + \delta], \quad (3.71)$$

where the constant k has been absorbed into the constant A. From Equation 3.59, the amplitude function $\beta(s)$ must satisfy the differential equation

$$2\beta\beta'' - \beta'^2 + 4\beta^2 K = 4. \quad (3.72)$$

It is often easier to remember this equation when written in terms of the Courant-Snyder parameters:

$$K\beta = \gamma + \alpha'. \quad (3.73)$$

The matrix of Equation 3.69 is often written in a compact way as

$$M = I \cos \Delta\psi_C + J \sin \Delta\psi_C, \quad (3.74)$$

where

$$J \equiv \begin{pmatrix} \alpha & \beta \\ -\gamma & -\alpha \end{pmatrix}. \quad (3.75)$$

Noting that $J^2 = -I$, where I is the identity matrix, one may also write Equation 3.74 in even more compact form:

$$M = e^{J\Delta\psi_C}. \quad (3.76)$$

The latter form often permits simplification of algebraic manipulations.

Computation of the Courant-Snyder parameters may be performed by comparing the two ways of expressing the matrix through a repeat period.

Suppose that multiplying all the individual matrices of the repeat period gives

$$M = \begin{pmatrix} a & b \\ c & d \end{pmatrix}. \tag{3.77}$$

Equating the two versions of M,

$$\begin{pmatrix} a & b \\ c & d \end{pmatrix} = \begin{pmatrix} \cos \Delta\psi_C + \alpha \sin \Delta\psi_C & \beta \sin \Delta\psi_C \\ -\gamma \sin \Delta\psi_C & \cos \Delta\psi_C - \alpha \sin \Delta\psi_C \end{pmatrix}. \tag{3.78}$$

Then, first of all,

$$\cos \Delta\psi_C = \tfrac{1}{2}(a + d) = \tfrac{1}{2} \operatorname{Tr} M. \tag{3.79}$$

Comparison of this relation with the identical one satisfied by μ in the stability discussion of Section 3.1.3 enables us to identify μ as the phase advance through a repeat period.

Knowing $\cos \Delta\psi_C$ gives us the magnitude but not the sign of $\sin \Delta\psi_C$. But β must be a positive quantity, so the sign of $\sin \Delta\psi_C$ is whatever the sign of the matrix element b happens to be. Then

$$\beta = \frac{b}{\sin \Delta\psi_C}, \tag{3.80}$$

and by subtraction of the diagonal elements

$$\alpha = \frac{a - d}{2 \sin \Delta\psi_C}. \tag{3.81}$$

Thus, we have the Courant-Snyder parameters at one point of the periodic lattice. But the same procedure works between any pair of corresponding points in the lattice, so one can find $\beta(s)$ for all s. With β determined at all points of the lattice, the particle motion from one point to another can be described by the matrix equation

$$\begin{pmatrix} x \\ x' \end{pmatrix}_{s_2} = M(s_1 \rightarrow s_2)\begin{pmatrix} x \\ x' \end{pmatrix}_{s_1}. \tag{3.82}$$

We can arrive at an explicit representation of the matrix $M(s_1 \rightarrow s_2)$ in terms of the amplitude function through the use of Equations 3.60 and 3.61. Suppose x_1 and x'_1 are the initial conditions at $s = s_1$. Then the constants A_1 and A_2 are

$$A_1 = \frac{x_1}{w_1}, \tag{3.83}$$

$$A_2 = \frac{x'_1 w_1 - x_1 w'_1}{k}. \tag{3.84}$$

Inserting these values for A_1 and A_2 into Equations 3.60 and 3.61 and rewriting in terms of the amplitude functions, the matrix $M(s_1 \to s_2)$ may be written as

$$
\begin{pmatrix}
\left(\dfrac{\beta_2}{\beta_1}\right)^{1/2} (\cos \Delta\psi + \alpha_1 \sin \Delta\psi) & (\beta_1\beta_2)^{1/2} \sin \Delta\psi \\[3mm]
-\dfrac{1 + \alpha_1\alpha_2}{(\beta_1\beta_2)^{1/2}} \sin \Delta\psi + \dfrac{\alpha_1 - \alpha_2}{(\beta_1\beta_2)^{1/2}} \cos \Delta\psi & \left(\dfrac{\beta_1}{\beta_2}\right)^{1/2} (\cos \Delta\psi - \alpha_2 \sin \Delta\psi)
\end{pmatrix}.
$$

$$(3.85)$$

Here, $\Delta\psi$ is the phase advance from s_1 to s_2.

The phase advance between any two points now can be uniquely determined via

$$
\Delta\psi(s_1 \to s_2) = \int_{s_1}^{s_2} \frac{ds}{\beta(s)}.
\tag{3.86}
$$

In particular, for a circular machine, the number of oscillations per turn,

$$
\nu \equiv \frac{1}{2\pi} \oint \frac{ds}{\beta(s)},
\tag{3.87}
$$

is called the *tune* of the accelerator. Since β can be interpreted as an oscillation's local wavelength divided by 2π, a sense of scale for the values of β is obtained. While the actual particle oscillations might have small amplitudes (e.g., millimeters), the amplitude function β should be expected to take on numerical values of the scale of the repeat period. This also tells us that the numerical value of the constant A describing a particle's motion will be comparatively rather small. Note that A has dimensions of (length)$^{1/2}$, and β has dimensions of length.

It should be pointed out that the solution to the equation of motion, Equation 3.71, explicitly implies stable motion. The solution must also be able to describe unstable motion. Demonstration of how the amplitude function and phase advance are altered for the case of an unstable lattice is left to the problems at the end of the chapter.

We have just gone through rather a lot of algebra to develop a way of representing a betatron oscillation that is at first sight a good deal more complicated than propagation using elementary matrices. In the thin lens approximation—not a bad approximation for large separated function synchrotrons—the betatron oscillation is after all just a sequence of straight line segments. We've managed to express these line segments in harmonic oscillator language; presumably there is some benefit. We'll try to illustrate the advantages as time goes on. Let us just point out that any oscillation can be easily constructed once one has a tabulation of the Courant-Snyder parameters and the phase advance as a function of position.

3.2.4 Emittance and Admittance

Now we are in a position to approach the important questions of the space demanded by the beam and the space provided by the accelerator. We are still working in the context of a perfect accelerator—no field imperfections.

In our solution for a betatron oscillation,

$$x(s) = A\sqrt{\beta(s)}\,\cos[\psi(s) + \delta], \tag{3.88}$$

the constant A can be expressed in terms of x and x' by eliminating the trigonometric functions. Forming the combination

$$\alpha(s)x(s) + \beta(s)x'(s) = -A\sqrt{\beta(s)}\,\sin[\psi(s) + \delta], \tag{3.89}$$

then squaring and summing Equations 3.88 and 3.89, we obtain

$$A^2 = \gamma(s)x(s)^2 + 2\alpha(s)x(s)x'(s) + \beta(s)x'(s)^2. \tag{3.90}$$

This Courant-Snyder invariant is analogous to the total energy of a harmonic oscillator. At any point in the accelerator, the invariant form describes an ellipse as depicted in Figure 3.11.

For the case of a circular accelerator, each time that the particle passes a particular position in the ring, its betatron oscillation coordinates will appear as a point on the ellipse given by the amplitude function and its slope at that point, as sketched in Figure 3.12.

For different locations through the lattice, the ellipses will have different shapes and orientations, but they will all have the same value of A. This

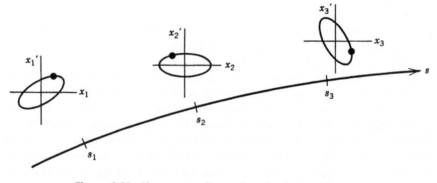

Figure 3.11. Phase space ellipses along the design trajectory.

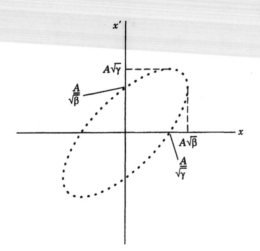

Figure 3.12. Phase space mapping from turn to turn in a circular accelerator.

means that they all have the same area. For, consider the general equation of an ellipse:

$$ax^2 + 2bxy + cy^2 = d. \tag{3.91}$$

According to analytic geometry, the area of this ellipse is

$$\frac{\pi d}{\sqrt{ac - b^2}}, \tag{3.92}$$

which in our case becomes

$$\frac{\pi A^2}{\sqrt{\beta\gamma - \alpha^2}} = \pi A^2. \tag{3.93}$$

The coordinates x, x' define a phase space for the motion that we are discussing, and we have shown that the area in this phase space enclosed by an unaccelerated particle is constant. Only a slight modification is necessary to find the invariant appropriate for accelerated motion and this is the subject of the next subsection.

The *admittance* is the phase space area associated with the largest ellipse that the accelerator will accept. From the preceding discussion, we would estimate the admittance as follows. At any point in the accelerator, the maximum value of x is $A\sqrt{\beta}$. If the half aperture available to the beam is $a(s)$, then somewhere there will be a minimum in $a(s)/\sqrt{\beta(s)}$. Then the

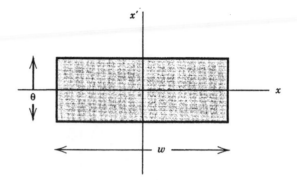

Figure 3.13. Phase space of particles emanating from a source of width w and angular spread θ.

admittance will be

$$\left(\pi \frac{a^2}{\beta} \right)_{\min}. \tag{3.94}$$

In the special case of a uniform half aperture a with no intruding septa, electrodes, and so on, then the minimum in $a/\sqrt{\beta}$ would occur at the maximum value of the amplitude function, $\beta_{\max}$; then

$$\text{admittance} = \frac{\pi a^2}{\beta_{\max}}. \tag{3.95}$$

The phase space area occupied by the beam is called the *emittance*, and is frequently denoted by ϵ. The ideal beam, of course, would have zero cross-sectional area and all the particles would be headed in exactly the same direction. All the particles would occupy the same point in the phase space of this transverse degree of freedom. But the most elementary model of a particle source leads to a nonzero emittance. Suppose that you have a source of width w from each point of which particles are produced within an angle θ. The phase space plot for the beam at the source will look like that depicted in Figure 3.13, enclosing a phase space area $w\theta$.

The phase space distribution of a beam is certainly not a uniformly populated rectangle, so a general definition of emittance will take this circumstance into account. For practical purposes, the phase space boundary of the beam may be considered to be an ellipse. Suppose that an irregularly shaped area is injected into an accelerator, with an initial state as shown in Figure 3.14. The subsequent motion of individual particles will lie on the elliptical invariant curves, as shown. As a result, the phase space demanded by the beam will be the area of the dashed curve. As time progresses, the

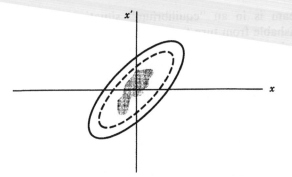

Figure 3.14. A beam with an irregularly shaped phase space distribution will conform to the elliptical phase space dictated by the optical properties of the accelerator, as a result of nonlinear forces inevitably present in a real machine.

initial phase space will tend to smear out and conform to the shape characteristic of the accelerator lattice as a result of field nonlinearities. In the static case that we are considering, the area in phase space remains constant (according to Liouville's theorem) but is so distorted that the area has increased in effect. This process of phase space dilution is called filamentation, and is avoided, insofar as is possible, by matching the injected beam shape to that of the invariant contours provided by the accelerator lattice. Emittance dilution will be examined in Chapter 7.

To summarize, then, if a beam in a synchrotron has emittance ϵ, then the phase space area is bounded by a curve

$$\frac{\epsilon}{\pi} = \gamma x^2 + 2\alpha xx' + \beta x'^2. \tag{3.96}$$

It is often convenient to speak of the emittance for a particular particle distribution in terms of the rms transverse beam size. As an example, we will consider a beam in a synchrotron in which the particles follow a Gaussian distribution in one transverse degree of freedom. This is the natural choice for electron storage rings, since synchrotron radiation ensures that the equilibrium distributions are Gaussian provided that particle losses are insignificant. The Gaussian distribution function is also a reasonable approximation for proton beams as well, in most instances.

Suppose that the distribution in the transverse coordinate x normalized to one particle is given by the density function

$$n(x)\,dx = \frac{1}{\sqrt{2\pi}\,\sigma}e^{-x^2/2\sigma^2}\,dx, \tag{3.97}$$

and that the distribution is stationary in time at a particular location s. That is, the beam is in an "equilibrium" situation where the distribution is indistinguishable from turn to turn. Since trajectories in x-$(\alpha x + \beta x')$ phase space are circles, as can be seen by examining Equations 3.88 and 3.89, then in this equilibrium situation the distribution in the coordinate $\alpha x + \beta x'$ will also be Gaussian with standard deviation σ. For, after all, the population on a given circle just rotates through an angle corresponding to the phase advance from turn to turn, and so the equilibrium distribution is independant of position along a circle, and depends only on the radius of the circle. The two-dimensional phase space distribution in these coordinates will be

$$n(x, \alpha x + \beta x') \, dx \, d(\alpha x + \beta x')$$

$$= \frac{1}{2\pi\sigma^2} e^{-[x^2 + (\alpha x + \beta x')^2]/(2\sigma)^2} \, dx \, d(\alpha x + \beta x'). \qquad (3.98)$$

We switch to polar coordinates, where the radial coordinate is

$$r^2 = x^2 + (\alpha x + \beta x')^2, \qquad (3.99)$$

and then the distribution is

$$n(r, \theta) r \, dr \, d\theta = \frac{1}{2\pi\sigma^2} e^{-r^2/2\sigma^2} r \, dr \, d\theta. \qquad (3.100)$$

If we define a radius a within which a fraction F of the particles are contained, then

$$F = \int_0^{2\pi} \int_0^a nr \, dr \, d\theta = \int_0^a e^{-r^2/2\sigma^2} \frac{r \, dr}{\sigma^2}, \qquad (3.101)$$

and, solving for a,

$$a^2 = -2\sigma^2 \ln(1 - F). \qquad (3.102)$$

Multiplying Equation 3.96 by β, we see that

$$\beta\epsilon/\pi = x^2 + (\alpha x + \beta x')^2. \qquad (3.103)$$

If this emittance is the area in x-x' phase space that contains the fraction F of the particles, then $\beta\epsilon/\pi = a^2$. Thus,

$$\pi a^2 = \beta\epsilon = -2\pi\sigma^2 \ln(1 - F), \qquad (3.104)$$

or

$$\epsilon = -\frac{2\pi\sigma^2}{\beta} \ln(1 - F). \qquad (3.105)$$

Table 3.1 The fraction *F* of a Gaussian beam
associated with various commonly used
definitions of the emittance

ϵ	F (%)
σ^2/β	15
$\pi\sigma^2/\beta$	39
$4\pi\sigma^2/\beta$	87
$6\pi\sigma^2/\beta$	95

Equation 3.105 gives the area in *x-x'* phase space which contains the fraction *F* of a Gaussian beam with transverse rms beam size σ at a point in the lattice where the amplitude function is β. Various authors and institutions make different choices for the fraction *F*, and so there is not a standard significance to the numbers quoted for emittance. Table 3.1 lists a number of common definitions of emittance and their associated fractions *F*. The first entry is the near-standard choice in the electron accelerator community, where the quantity of primary interest is the rms beam size. The third and fourth entries tend to be used in proton accelerators in circumstances where one wishes to characterize the total beam size. For the remainder of the text we will primarily use the second entry, as it combines the significance of rms quantities with the emphasis on emittance as a phase space area.

The two most frequently used relations are those that give the maximum displacement and angle anywhere around the ring:

$$x_{\text{max}} = \sqrt{\frac{\epsilon\beta_{\text{max}}}{\pi}} \tag{3.106}$$

and

$$x'_{\text{max}} = \sqrt{\frac{\epsilon\gamma_{\text{max}}}{\pi}} . \tag{3.107}$$

Of course, the total number of particles contained within these maxima will depend on the choice of *F* in the relations above.

3.2.5 Adiabatic Damping of Betatron Oscillations

In the previous discussion we considered only the motion of particles with constant total momentum. We wish now to show how the amplitude of the motion varies as a function of the particle momentum, assuming that the momentum is a slowly changing function of time, or equivalently, of longitudinal position in the case of a linear accelerator. To illustrate the principle but avoid complexity, we begin with the equation of motion for a charged particle in the presence of a magnetic field of the form $\vec{B} = \hat{y}B'x$:

$$\frac{d}{dt}(p_x) = (e\vec{v} \times \vec{B})_x = -ev_s B_y = -evB'x. \tag{3.108}$$

Noting that $p_x = px'$,

$$v\frac{d}{ds}(px') = v(px'' + p'x') = -evB'x, \tag{3.109}$$

or

$$x'' + \frac{p'}{p}x' + \frac{eB'}{p}x = 0. \tag{3.110}$$

This is just Hill's equation with an added damping term.

To solve the above differential equation, we assume the solution to be of the form $x = uv$. Then

$$vu'' + \left(2v' + \frac{p'}{p}v\right)u' + \left(v'' + \frac{p'}{p}v' + \frac{eB'}{p}v\right)u = 0. \tag{3.111}$$

We now choose the function v such that the u' term is zero. That is,

$$2\frac{v'}{v} = -\frac{p'}{p}, \tag{3.112}$$

which says that v is of the form

$$v = v_0\left(\frac{p_0}{p}\right)^{1/2}. \tag{3.113}$$

Since the momentum is changing slowly, the v'' and $p'v'$ terms may be neglected. To see this, consider the coefficient of u in the differential equation above. With the form of v just obtained, this coefficient is

$$v_0\left(\frac{p_0}{p}\right)^{1/2}\left[\frac{eB'}{p} - \frac{1}{4}\left(\frac{p'}{p}\right)^2 - \frac{1}{2}\frac{p''}{p}\right]. \tag{3.114}$$

Since p is changing slowly, p'' is negligible with respect to the p'^2 term. Now the second term is on the order of

$$\left(\frac{p'}{p}\right)^2 = \left(\frac{\Delta p}{2\pi R}\frac{1}{p}\right)^2 = \frac{1}{4\pi^2 R^2}\left(\frac{\Delta p}{p}\right)^2, \tag{3.115}$$

where Δp, the momentum increase per passage of the accelerating stations, is typically very much smaller than p. Since this term is much smaller than the centripetal term $\sim 1/\rho^2$ which we have already neglected from the outset of this discussion, the differential equation reduces to

$$\left(u'' + \frac{eB'}{p}u \right)v = 0, \tag{3.116}$$

or,

$$u'' + \frac{eB'}{p}u = 0, \tag{3.117}$$

which is Hill's equation. Therefore, the complete solution is of the form

$$x = uv = A_0 \left(\frac{p_0}{p} \right)^{1/2} \beta^{1/2}(s)\cos[\psi(s) + \delta]. \tag{3.118}$$

The amplitude of the betatron oscillation is thus damped as the energy of the beam is adiabatically increased. Since the beam emittance is defined as a phase space area bounded by a Courant-Snyder invariant curve, and since this area is proportional to the square of the betatron amplitude, as shown earlier, we see that the beam emittance varies inversely with the beam momentum. The use of a *normalized* emittance,

$$\epsilon_N \equiv \epsilon \times \left(\gamma \frac{v}{c} \right), \tag{3.119}$$

permits comparisons of phase space areas independent of kinematic factors. This normalized emittance should, in the ideal world, remain constant throughout the entire acceleration process. The fact that it does not is exemplified in Chapters 7 and 8.

3.3 MOMENTUM DISPERSION

We have just examined the motion of particles having the same momentum as the ideal particle but differing transverse position and direction. Now we want to study the motion of particles differing in momentum from that of the ideal particle. The source of such differences is the bend fields that establish the ideal trajectory. This is a moot point for the case of a linear accelerator, but for circular accelerators it is a primary design consideration. We will find that these *off-momentum* particles in general undergo betatron oscillations about a new class of closed orbits in circular accelerators. The displacement of these closed orbits from that of the ideal particle will be described by a new lattice function—the momentum dispersion function.

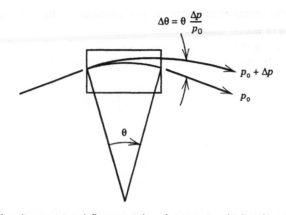

Figure 3.15. A bending magnet deflects particles of momentum higher than that of the ideal particle through a lesser angle, leading to a variety of closed orbits for particles of differing momenta.

The momentum dispersion function has its origin in the simple fact that a particle of higher momentum is deflected through a lesser angle in a bending magnet, as illustrated in Figure 3.15. It is as though a bending magnet which deflects the ideal particle by θ is the source of an angular perturbation $\theta \Delta p/p$ to the trajectory of an off-momentum particle entering on the design orbit, where $\Delta p/p$ is the fractional momentum difference relative to the ideal momentum.

In addition, higher momentum particles are bent less effectively in the focusing elements. That is, there is an effect completely analogous to chromatic aberration in conventional optics. The dependence of focusing on momentum will bring in its wake a dependence of betatron oscillation tune on momentum; the parameter quantying this relationship is called the chromaticity.

The compensation of chromaticity is accomplished by sextupole magnets, and we are led to the introduction of nonlinear elements into our heretofore linear focusing structure.

3.3.1 Equation of Motion for an Off-Momentum Particle

Since all of the treatment of the previous two sections was based on particles having the same momentum, we must begin again with the equation of motion to find the orbits for particles of differing momenta. We start from Equation 3.42, where we had

$$x'' - \frac{\rho + x}{\rho^2} = -\frac{B_y}{(B\rho)}\left(1 + \frac{x}{\rho}\right)^2. \tag{3.120}$$

Recalling that the magnetic rigidity is the momentum per unit charge, we will let $(B\rho)$ represent the magnetic rigidity of the ideal particle with momentum p_0 and include a factor of p_0/p on the right hand side of the above equation of motion. That is,

$$x'' - \frac{\rho + x}{\rho^2} = -\frac{B_y}{(B\rho)}\left(1 + \frac{x}{\rho}\right)^2 \frac{p_0}{p}. \qquad (3.121)$$

Furthermore, we still consider fields which vary linearly with transverse position, i.e.,

$$B_y = B_0 + B'x, \qquad (3.122)$$

and expand the above equation of motion, neglecting terms quadratic in x/ρ and higher. We get

$$x'' + \left\{\frac{1}{\rho^2}\frac{2p_0 - p}{p} + \frac{B'}{(B\rho)}\frac{p_0}{p}\right\}x = \frac{1}{\rho}\frac{\Delta p}{p}, \qquad (3.123)$$

where $\Delta p \equiv p - p_0$.

Let us write the closed orbit of an off-momentum particle in the form

$$x = D(p,s)\frac{\Delta p}{p_0} \qquad (3.124)$$

and then look for the closed solution. That is, we look for the solution subject to the condition

$$D(p, s + C) = D(p, s), \qquad (3.125)$$

where, as before, C is the repeat distance of the hardware. This function is referred to as the dispersion function. Since Equation 3.124 is a solution to the inhomogeneous Hill's equation, the general solution will differ from this particular (closed) solution by the addition of a solution of the homogeneous equation.

The equation to be solved for D is thus

$$D'' + \left\{\frac{1}{\rho^2}\frac{2p_0 - p}{p} + \frac{B'}{(B\rho)}\frac{p_0}{p}\right\}D = \frac{1}{\rho}\frac{p_0}{p}. \qquad (3.126)$$

This is an inhomogeneous Hill's equation. The right hand side indicates that bending is the source of momentum dispersion. A bend center increments the slope of the dispersion function by the bending angle at the momentum in question, just as suggested by Figure 3.15. Since betatron oscillation wavelengths are apt to be long compared to the lengths of bending magnets, it is a good approximation to say that such a magnet changes the slope of the dispersion function by the magnet's bend angle at its bend center. (For a pure dipole magnet, this is an exact statement.)

3.3.2 Solution of Equation of Motion

We can proceed in the same fashion as in the "piecewise" method used for the betatron oscillations. Our equation is

$$D'' + K(s)D = \frac{1}{\rho}\frac{p_0}{p} = \frac{eB_0(s)}{p}, \tag{3.127}$$

where K is taken to be constant in each element of the lattice. Let us also assume B_0 is a constant within each element.

We can adapt the matrix approach used in the discussion of betatron oscillations to the present case. Since a particular solution to the dispersion function equation is a constant in each element, it may be readily shown that the general solution may be written as

$$\begin{pmatrix} D \\ D' \end{pmatrix}_{out} = \begin{pmatrix} a & b \\ c & d \end{pmatrix}\begin{pmatrix} D \\ D' \end{pmatrix}_{in} + \begin{pmatrix} e \\ f \end{pmatrix}, \tag{3.128}$$

where the two-by-two matrix is the same as that for the treatment of betatron oscillations (the homogeneous solution). By the addition of a third trivial equation, $1 = 1$, the equation above can be expressed in terms of a three-by-three matrix:

$$\begin{pmatrix} D \\ D' \\ 1 \end{pmatrix}_{out} = \begin{pmatrix} a & b & e \\ c & d & f \\ 0 & 0 & 1 \end{pmatrix}\begin{pmatrix} D \\ D' \\ 1 \end{pmatrix}_{in}. \tag{3.129}$$

Table 3.2 shows the values of the matrix elements e and f for the cases where $K < 0$, $K = 0$, and $K > 0$.

By multiplying the various matrices for the pieces of the ring, we can find a matrix M for one turn (or, for one repeat period). The condition for the displaced equilibrium orbit to be closed is

$$\begin{pmatrix} D \\ D' \\ 1 \end{pmatrix} = M\begin{pmatrix} D \\ D' \\ 1 \end{pmatrix}, \tag{3.130}$$

and its solution yields the dispersion function at the starting point.

Table 3.2 Values of the matrix elements M_{13} and M_{23} for various ranges of the spring constant K

K	e	f
< 0	$\dfrac{e}{p\lvert K\rvert}B_0(\cosh\sqrt{\lvert K\rvert}\,l - 1)$	$\dfrac{e}{p\sqrt{\lvert k\rvert}}B_0\sinh\sqrt{\lvert K\rvert}\,l$
0	$\dfrac{1}{2}\dfrac{eB_0 l}{p}l$	$\dfrac{eB_0 l}{p}$
> 0	$\dfrac{e}{pK}B_0(1 - \cos\sqrt{K}\,l)$	$\dfrac{e}{p\sqrt{K}}B_0\sin\sqrt{K}\,l$

Note that the same 3×3 matrices propagate either the dispersion function or the trajectory itself. That is, M operates on the vector

$$\begin{pmatrix} D \\ D' \\ 1 \end{pmatrix} \tag{3.131}$$

or on the vector

$$\begin{pmatrix} x \\ x' \\ \Delta p \\ p_0 \end{pmatrix}. \tag{3.132}$$

The same procedure can be carried out starting at any point in the lattice, and so the dispersion function can be computed everywhere. Alternatively, having found the dispersion function and its derivative at a particular starting point, this solution can be propagated forward using either the differential equation or the matrices describing individual elements.

In simple situations, the dispersion function will be everywhere positive; that is, orbits of higher momenta than the design orbit are at larger radius. The difference in perimeter between off-momentum orbits and the design orbit is characterized by the *compaction factor*, which is unfortunately designated by the overworked symbol α and is defined by the relation

$$\frac{\Delta C}{C} = \alpha\frac{\Delta p}{p_0}, \tag{3.133}$$

where C is the accelerator circumference. The form of this equation suggests that the name stems from the circumstance that this parameter is less than unity. That is indeed the case.

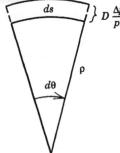

Figure 3.16. Illustration of increment in path length difference between off-momentum particle and ideal particle.

To illustrate the calculation of the compaction factor, let us suppose that the bending is provided by sector magnets. A *sector magnet* is one in which the ideal orbit enters and exits at right angles to the magnet ends. The procedure for calculating the difference in path between a particle of the central momentum and an off-momentum particle is sketched in Figure 3.16. The circumference change, ΔC, is given by

$$\Delta C = \oint \left(\rho + D \frac{\Delta p}{p_0} \right) d\theta - \oint (\rho \, d\theta). \tag{3.134}$$

For the fractional circumference change we therefore have

$$\frac{\Delta C}{C} = \frac{\oint D \, ds/\rho}{\oint ds} \frac{\Delta p}{p_0} = \left\langle \frac{D}{\rho} \right\rangle \frac{\Delta p}{p_0}, \tag{3.135}$$

or

$$\alpha \equiv \frac{1}{\gamma_t^2} = \left\langle \frac{D}{\rho} \right\rangle. \tag{3.136}$$

For a simple lattice, $\gamma_t \approx$ tune, as is the subject of a problem at the end of this chapter. The tune of an alternating gradient lattice scales with the number of cells, and so γ_t increases at the designer's discretion with the size of the accelerator. So again we have a circumstance where an aperture requirement—in this case, the aperture set aside to accommodate the momentum spread—is decoupled from the overall scale of the accelerator. This is another major advantage of the strong focusing principle.

We can build on our discussion of emittance to characterize the total beam size due to both betatron oscillations and momentum spread. For, we can write the displacement from the ideal trajectory of a particle as the sum of two terms:

$$x = D \frac{\Delta p}{p_0} + x_\beta, \qquad (3.137)$$

where the first term represents the contribution of the closed orbit of the off-momentum particle and the second the free oscillation about that closed orbit. Averaging the square of this expression yields for the rms displacement

$$\sigma^2 = \frac{\epsilon \beta}{\pi} + D^2 \left\langle \left(\frac{\Delta p}{p_0} \right)^2 \right\rangle, \qquad (3.138)$$

and we leave it to the reader to determine which of the various definitions of transverse emittance has been used to arrive at this last expression.

3.4 LINEAR DEVIATIONS FROM THE IDEAL LATTICE

We have now brought to an end our discussion of the linear transverse dynamics of the ideal accelerator. Thus far we have established the basic principles underlying single particle accelerator design. This is the framework upon which we will build in future chapters. We conclude this chapter with an introductory treatment of linear deviations from the ideal lattice, including steering and tune errors and adjustments.

3.4.1 Steering Errors and Corrections

Thus far we have described an accelerator where each magnetic or electric element performs exactly its prescribed task. In particular, for a circular accelerator, there is by design an ideal reference orbit which closes on itself and about which betatron oscillations would occur for particles with nonzero emittance. Suppose, however, that a particular bending magnet has a field somewhat different from its intended value. We will show that the field imperfection leads to a new closed orbit near, but differing from, the ideal design orbit.

Suppose that in our otherwise ideal circular accelerator a single steering error of magnitude $\theta = \Delta B l / (B\rho)$ is located at $s = 0$, where ΔB is an unintentional uniform field over the path length l. We want to find the new closed orbit, for x identically zero is no longer a solution of the equation of motion. Everywhere except at the location of the steering error, however, the equation of motion is still that of a betatron oscillation—only at the point of the angular impulse is the equation inhomogeneous.

Let the orbit immediately downstream of the deflection θ be specied by x_0, x_0'. To propagate this initial condition around the ring, we just multiply by the single-turn matrix M. Now we are immediately upstream of the deflection; to close the orbit we need only add the angle θ and demand that we be back to x_0, x_0'. In symbols,

$$M\begin{pmatrix} x_0 \\ x_0' \end{pmatrix} + \begin{pmatrix} 0 \\ \theta \end{pmatrix} = \begin{pmatrix} x_0 \\ x_0' \end{pmatrix}. \tag{3.139}$$

Solving for x_0 and x_0',

$$\begin{pmatrix} x_0 \\ x_0' \end{pmatrix} = (I - M)^{-1}\begin{pmatrix} 0 \\ \theta \end{pmatrix}. \tag{3.140}$$

The matrix $(I - M)^{-1}$ can be recast using Equation 3.76. Then

$$(I - M)^{-1} = (I - e^{J2\pi\nu})^{-1} = \left[e^{J\pi\nu}(e^{-J\pi\nu} - e^{J\pi\nu})\right]^{-1} \tag{4.141}$$

$$= -(2J\sin\pi\nu)^{-1}(e^{J\pi\nu})^{-1} \tag{3.142}$$

$$= \frac{1}{2\sin\pi\nu}Je^{-J\pi\nu} \tag{3.143}$$

$$= \frac{1}{2\sin\pi\nu}(J\cos\pi\nu + I\sin\pi\nu), \tag{4.144}$$

and the closed orbit at $s = 0$ will be

$$\begin{pmatrix} x_0 \\ x_0' \end{pmatrix} = (I - M)^{-1}\begin{pmatrix} 0 \\ \theta \end{pmatrix} = \frac{\theta}{2\sin\pi\nu}\begin{pmatrix} \beta_0\cos\pi\nu \\ \sin\pi\nu - \alpha_0\cos\pi\nu \end{pmatrix}. \tag{3.145}$$

The closed orbit may now be expressed as a function of position or phase around the ring by, for instance, applying the matrix for propagation between one point and another (Equation 3.85). After a small bit of algebra, one finds

$$x(s) = \frac{\theta\beta^{1/2}(s)\beta_0^{1/2}}{2\sin\pi\nu}\cos[\psi(s) - \pi\nu] \tag{3.146}$$

for $0 < \psi < 2\pi\nu$.

The solution for the closed orbit is sketched in Figure 3.17. We see that the new closed orbit in the presence of a single steering error exhibits a *cusp* at the location of the error, where the trajectory experiences a *kink* through an angle θ. A local maximum in the oscillation occurs at the point in the accelerator where $\psi = \pi\nu$, i.e., half way around the accelerator for typical lattice designs. Particles whose initial conditions do not coincide with this closed orbit will still be influenced by the steering error each turn and will, in fact, perform betatron oscillations about the new closed orbit.

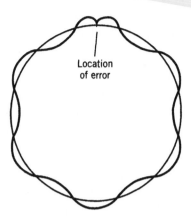

Figure 3.17. Sketch of closed orbit in presence of a single steering error.

In reality there will be as many steering errors as there are magnetic components in the accelerator. Let us continue by looking at the case of a simple FODO lattice in a synchrotron. Ideally, bending is provided by dipole magnets having a constant magnetic field in their transverse coordinates, while focusing is provided by quadrupole lenses. Most present day high energy synchrotrons employ this *separated function* lattice or a closely related variant. The dipole magnets will produce steering errors in both transverse degrees of freedom due to construction errors as well as installation and alignment errors. Variations in the total vertical component of the field as seen along the ideal trajectory will cause deflections in the horizontal plane as the particle passes through the magnet. Likewise, horizontal field components, such as those present when a dipole magnet is rotated slightly about its longitudinal axis, will cause vertical steering errors. A transverse alignment error of a quadrupole magnet will place a transverse field component on the ideal trajectory and so also contribute to the sources of steering errors.

As a numerical example, consider a quadrupole magnet which is displaced from the ideal trajectory by an amount δ in one of the transverse coordinates. Then, as a particle passes through this element, the slope of its trajectory will change by an amount $\Delta x' = \delta/F \equiv \theta$, where F is the focal length of the quadrupole. Take the case of the Tevatron, where the standard quadrupole focal length is 25 m. Then, an alignment error of 0.5 mm will generate an angular steering error of $\theta = 20~\mu\text{rad}$. If the error is at a focusing quadrupole where the amplitude function is 100 m, and one takes the tune to be about 19.4, then Equation 3.146 yields for a maximum closed orbit distortion at the focusing quadrupoles a value of

$$\Delta \hat{x} = \left| \frac{(20~\mu\text{rad})(100~\text{m})}{2 \sin \pi (19.4)} \right| = 1~\text{mm}.$$

This is a linear problem, so we may use the superposition principle to combine the effects of many such errors. If there are N errors in the accelerator located at equivalent values of the amplitude function, we would expect the rms value of the closed orbit distortion to be larger than the preceding figure by a factor of order $N^{1/2}$ with the placement error δ also reinterpreted as an rms value. In the Tevatron, $N \approx 100$, so we come to an estimate of 10 mm for the rms orbit distortion due to quadrupole placement errors, and we would expect peaks larger than the rms by some factor on the order of 3 or 4, depending upon the distribution. As a consequence, we need some means of correcting steering errors as a basic design feature in synchrotrons of this scale.

Correction of closed orbit distortions can be carried out with the aid of a set of independently powered steering dipole magnets. The same set of steering dipoles can be used to make intentional closed orbit distortions to facilitate a variety of accelerator functions. More rigorous and quantitative calculations of the above are the subject of some of the problems at the end of the chapter.

As a final note, referring to Equation 3.146, we observe that there is no closed orbit if the tune is an integer. This is the most elementary example of a resonance. Of course, we didn't need to go through any algebra to find that out. If the tune were an integer, the steering errors would just reinforce from turn to turn until the oscillation amplitude became large enough to strike the walls of the vacuum chamber. The implication in the formula that the orbit goes to infinity is just an artifact of our approximations. But since infinity is only a few centimeters away, the approximations are pretty good.

3.4.2 Focusing Errors and Corrections

A gradient error would be expected to alter the tune of a circular accelerator. Let there be a single gradient error equivalent to a thin lens quadrupole with focal length f. The matrix M for a single turn is then

$$M = M_0 \begin{pmatrix} 1 & 0 \\ -\dfrac{1}{f} & 1 \end{pmatrix}, \tag{3.147}$$

where M_0 is the matrix for the ideal ring. From the trace of M it follows that

$$\cos 2\pi\nu = \cos 2\pi\nu_0 - \frac{1}{2}\frac{\beta_0}{f}\sin 2\pi\nu_0, \tag{3.148}$$

where ν and ν_0 are the new and old tunes respectively, and β_0 is the original amplitude function at the point of the perturbation. For the ideal ring, presumably ν_0 is real by design. But depending on the sign and magnitude of the gradient error term, ν can become complex; that is, the motion can become unstable. Since, for small magnitudes of the gradient error term, the

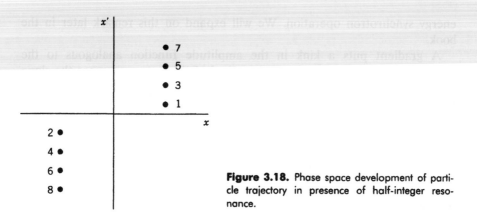

Figure 3.18. Phase space development of particle trajectory in presence of half-integer resonance.

instability will occur for ν near an integer or half integer, these instabilities are called half-integer resonances. There will be a range of values of ν_0 for which the motion is unstable; this range is called a *stopband*.

Just as in the case of dipole error resonances, we didn't need to use any algebra to demonstrate that quadrupole errors can produce resonance effects. Figure 3.18 represents the phase space history of a particle on successive turns as it passes the gradient error. The initial motion, in the absence of the error, was one in which the tune was an odd multiple of one-half. Successive passages of the gradient error just add constant vectors parallel to the vertical axis.

If the tune is not near a half-integer and the perturbation is sufficiently small, we can obtain a useful expression for the tune shift due to a gradient error by writing

$$\nu = \nu_0 + \delta\nu \tag{3.149}$$

and expanding the cosine term on the left hand side of the last equation. The result is

$$\delta\nu = \frac{1}{4\pi}\frac{\beta_0}{f}. \tag{3.150}$$

If there is a distribution of gradient errors, this last result generalizes to

$$\delta\nu = \frac{1}{4\pi}\sum_i \frac{\beta_i}{f_i} \rightarrow \frac{1}{4\pi}\oint \frac{\beta(s)B'(s)}{(B\rho)}\,ds \tag{3.151}$$

and is the lowest order (in gradient error) approximation to the tune shift.

In analogy with steering errors and corrections, one can make adjustments to the tunes of the accelerator by intentionally introducing perturbations on the gradients. The capability to adjust the tune is essential to modern high energy synchrotron operation. We will expand on this remark later in the book.

A gradient puts a kink in the amplitude function analogous to the deflection of the orbit produced by a dipole field. Let us compare the slope of the amplitude function on either side of the gradient error, or, equivalently, the parameter α, by calculating

$$\Delta M = \begin{pmatrix} 1 & 0 \\ -\dfrac{1}{f} & 1 \end{pmatrix} M_0 - M_0 \begin{pmatrix} 1 & 0 \\ -\dfrac{1}{f} & 1 \end{pmatrix}, \tag{3.152}$$

from which

$$\Delta \alpha = \frac{\beta}{f}, \tag{3.153}$$

where, in this case, β is the amplitude function of the perturbed lattice at the location of the error. Given the change in slope of the amplitude function produced by a gradient error, one can go on and calculate the deviation in the amplitude function throughout the ring. The principle of superposition is not valid for amplitude function perturbation due to combinations of gradient errors even in the linear lattice, in contrast to the situation for steering errors; but for sufficiently small gradient error, the treatments and results are quite similar. These matters are to be found in the problems at the end of the chapter.

3.4.3 Chromaticity

As mentioned during our treatment of the dispersion function, the equation of motion for an off-momentum particle includes a "spring constant" term which depends upon the particle momentum. One would thus suspect that the frequency of betatron oscillations of a particle, that is, the tune, would depend also upon its momentum. The change in tune due to momentum is called the chromaticity and is defined by

$$\delta \nu = \xi(p) \frac{\Delta p}{p_0}, \tag{3.154}$$

where $\Delta p / p_0$ is the momentum deviation relative to the ideal momentum.

The chromaticity can be calculated by Equation 3.150 relating tune shifts to gradient errors. Recall that we had

$$\delta \nu = \frac{1}{4\pi} \sum \frac{\beta_i}{f_i}, \tag{3.155}$$

where the f_i are to be identied as the focal lengths of the "error" quadrupoles that represent the difference between the off-momentum and central momentum states. That is, for a quadrupole of focal length F,

$$\frac{1}{f} = \frac{p_0}{p}\frac{1}{F} - \frac{1}{F} = -\frac{1}{F}\frac{\Delta p}{p},\tag{3.156}$$

and so

$$\xi = -\frac{1}{4\pi}\sum\frac{\beta_i}{F_i},\tag{3.157}$$

or, equivalently,

$$\xi = -\frac{1}{4\pi}\oint K\beta\,ds.\tag{3.158}$$

For a simple lattice, the chromaticity is about equal in magnitude and opposite in sign to the tune, as is to be demonstrated in one of the problems. In the more complex lattices of storage rings, the chromaticity is apt to be considerably larger in magnitude. This arises as follows. In order to focus the beams to a small spot at the collision point, one makes near-parallel beams on either side; that is, the amplitude function becomes large in quadrupoles which tend to be stronger than those elsewhere in the ring. The integrand in the last equation is therefore unusually large in such regions.

Another way of looking at the origin of the larger than normal chromaticity of storage rings is to make use of the differential equation satisfied by the amplitude function

$$K\beta = \alpha' + \gamma\tag{3.159}$$

to rewrite the expression for the chromaticity as

$$\xi = -\frac{1}{4\pi}\oint(\alpha' + \gamma)\,ds.\tag{3.160}$$

The first term vanishes when integrated around the ring. The term in γ will be large where the amplitude function is small, which is the situation at the collision point. Since γ is a constant in a drift space, the straight section in which collisions occur will contribute to the chromaticity a quantity equal to the straight section length divided by the amplitude function at the crossing for typical interaction region designs in which β is a minimum at the crossing point.

The source of chromaticity discussed here is the dependence of focusing strength on momentum for the ideal accelerator fields; the resulting chromaticity is called the *natural chromaticity*. There will be additional sources, as we shall see in Chapter 4.

Why worry about chromaticity? There are two reasons. If the beam has a large momentum spread, then a large chromaticity may place some portions of the beam on resonances. Secondly, the value of the chromaticity may determine whether or not certain intensity dependent motion is stable or unstable, as will be discussed in Chapter 6.

The capabilities for chromaticity adjustment provided by the linear lattice are limited; some other means must be used to modify this quantity. What is needed is a magnet that presents a gradient that is a function of momentum. A distribution of sextupole magnets is normally used for this purpose. In the horizontal plane, the sextupole field is of the form

$$B = kx^2, \tag{3.161}$$

and so the field gradient on a displaced equilibrium orbit is

$$B' = 2kx = 2kD\frac{\Delta p}{p_0}, \tag{3.162}$$

and the contribution to the chromaticity from sextupoles may be readily calculated.

A standard way of adjusting the chromaticity in both transverse degrees of freedom is to place a sextupole magnet at each main quadrupole location. The sextupoles are connected in two circuits; those at horizontally focusing quadrupoles are powered in one series circuit, and those at vertically focusing quadrupoles in the other. For the usual FODO lattice, the chromaticity changes due to the sextupoles would then be given by

$$\Delta\xi_H = \frac{N}{2\pi}(\beta_{max}D_{max}S_F + \beta_{min}D_{min}S_D), \tag{3.163}$$

$$\Delta\xi_V = -\frac{N}{2\pi}(\beta_{min}D_{max}S_F + \beta_{max}D_{min}S_D). \tag{3.164}$$

Here, N is the number of cells, and the sextupole strengths S_F and S_D are defined by $(\partial^2 B_y/\partial x^2) \cdot \text{length}/(2B\rho)$ evaluated at the focusing and defocusing quadrupole locations, respectively. Unfortunately, the sextupoles inevitably introduce intrinsically nonlinear aberrations, and we have taken our first step away from a design based upon purely linear dynamics.

PROBLEMS

1. In the "weak focusing" accelerators, the field index is defined as

$$n \equiv -\frac{dB/B}{dr/r}.$$

(a) Show that the equation of motion of a particle in the vertical (axial) degree of freedom is

$$\frac{d^2z}{dt^2} + \omega^2 nz = 0,$$

where ω is the angular rotation frequency. Thus, vertical oscillations are stable so long as $n > 0$.

(b) If the design radius for the machine is R and a particle's radial coordinate is $r = R + x$, where $x \ll R$, show that the equation of motion in the radial (horizontal) degree of freedom for small oscillations is

$$\frac{d^2x}{dt^2} + \omega^2(1 - n)x = 0.$$

Therefore, radial stability requires $n < 1$; stability in both transverse degrees of freedom simultaneously is assured only if $0 < n < 1$.

2. Consider a system made up of two thin lenses each of focal length f, one focusing and one defocusing, separated by a distance L. Show that the system is focusing if $|f| > L$.

3. Evaluate the matrix for a quadrupole of length 10 meters, with a gradient of 80 tesla/meter, and traversed by a particle with an energy of 20 TeV. Compare with a product of thin lens matrices occupying the same length.

4. Suppose that a particle traverses, first, a focusing lens with a focal length F; second, a drift of length L; third, a defocusing lens with focal length F; and, fourth, another drift of length L. Show that the matrix for this cell is given by

$$M = \begin{pmatrix} 1 - \dfrac{L}{F} - \left(\dfrac{L}{F}\right)^2 & 2L + \dfrac{L^2}{F} \\ -\dfrac{L}{F^2} & 1 + \dfrac{L}{F} \end{pmatrix}.$$

5. Consider a lattice made up entirely of thick gradient magnets, each of length L, alternating in "spring constant" between K and $-K$ (that is, assume that the $1/\rho^2$ term in the horizontal equation of motion is negligible). For what values of $\sqrt{K}L$ is the transverse motion in such a system stable?

6. Using a graphics terminal or home computer, and with $F = L$ in Problem 4 above, show that the motion is stable in transverse phase space. That is, start out particles with various values of x and x' and demonstrate that no divergence to large amplitude motion is indicated. Set $F = L/3$ and repeat; is the motion still stable?

7. Find the eigenvalues and eigenvectors for the FODO case. None is real —is that a problem?

8. Why is the transport matrix unimodular? Prove that it must be so under our assumptions. Suppose the particle energy changes—does the determinant still have to equal unity? If the matrix is

$$M = \begin{pmatrix} a & b \\ c & d \end{pmatrix}$$

show (easily) that the inverse is

$$M^{-1} = \begin{pmatrix} d & -b \\ -c & a \end{pmatrix}.$$

9. Suppose that one elects to write the matrix M for a periodic section as an exponential operator e^K. Using the properties of M as developed in Section 3.1.3, show that the trace of K is zero.

10. Going from point 1 to point 2, you traverse a sequence of elements that yield a matrix

$$M(1,2) = \begin{pmatrix} a & b \\ c & d \end{pmatrix}.$$

From point 2 to point 3, you traverse the same elements but in reverse order. Show that the matrix from 2 to 3 is

$$M(2,3) = \begin{pmatrix} d & b \\ c & a \end{pmatrix}.$$

11. Show that the amplitude function is a solution of the linear differential equation

$$\beta''' + 4\beta'K + 2\beta K' = 0.$$

Within a lattice element, where K is constant, the solution must be one of the three forms

$$\beta = a + bs + cs^2$$

$$= a \cos 2\sqrt{K}\, s + b \sin 2\sqrt{K}\, s + c$$

$$= a \cosh 2\sqrt{|K|}\, s + b \sinh 2\sqrt{|K|}\, s + c.$$

In each case evaluate a, b, c in terms of α_0 and β_0, the parameters at the beginning of the element.

12. Show that the maximum and minimum values of the amplitude function for the simple FODO cell are given by

$$\beta_{max} = 2F\left(\frac{1 + \sin(\mu/2)}{1 - \sin(\mu/2)}\right)^{1/2} = 2L\left(\frac{1 + \sin(\mu/2)}{\sin \mu}\right),$$

$$\beta_{min} = 2F\left(\frac{1 - \sin(\mu/2)}{1 + \sin(\mu/2)}\right)^{1/2} = 2L\left(\frac{1 - \sin(\mu/2)}{\sin \mu}\right).$$

Evaluate these for a quadrupole spacing of 100 m and phase advance per cell of 80°.

13. Suppose that a particle traveling along the design orbit experiences an angular deflection θ. Show that thereafter its motion is given by

$$x = \theta \beta_0^{1/2} \beta^{1/2}(s) \sin \psi,$$

where β_0 is the amplitude function at the point of deflection, and phase ψ is measured relative to that point. Calculate the oscillation amplitude associated with a 100 μrad deflection at a maximum β point in the lattice of the preceding problem.

14. Given the Courant-Snyder parameters, or equivalently the J-matrix, at one point in the ring, they may readily be found at other points with the use of the appropriate transfer matrices. Suppose J_1 is the matrix representing a known set of parameters, and we want to find J_2. Let $M(1, 2)$ be the matrix propagating the motion from point 1 to point 2. Show that the J-matrices at the two points are related by

$$J_2 = M(1, 2) J_1 M^{-1}(1, 2).$$

Show that the parameter relations are

$$\beta_2 = m_{11}^2 \beta_1 - 2m_{11}m_{12}\alpha_1 + m_{12}^2\gamma_1,$$

$$\alpha_2 = -m_{11}m_{21}\beta_1 + (m_{11}m_{22} + m_{12}m_{21})\alpha_1 - m_{12}m_{22}\gamma_1,$$

$$\gamma_2 = m_{21}^2\beta_1 - 2m_{21}m_{22}\alpha_1 + m_{22}^2\gamma_1,$$

where the m_{ij} are the matrix elements of $M(1, 2)$. We have used the relation between the elements of $M(1, 2)$ and the elements of the inverse in writing the above.

15. Show that the phase advance from point 1 to point 2 through a section described by the transport matrix $M(1, 2)$ is given by

$$\Delta\psi = \tan^{-1}\left(\frac{m_{12}}{\beta_1 m_{11} - \alpha_1 m_{12}}\right)$$

where the m_{ij} are the matrix elements of $M(1, 2)$ and β_1 and α_1 are the values of the amplitude functions at point 1.

16. For the simple thin lens FODO cell, verify that the phase advance as calculated by

$$\psi = \int \frac{ds}{\beta}$$

agrees with the result given by the trace of the matrix. Why is agreement assured in this case, while in general (for larger lattice segments) there may be ambiguities?

17. In order to achieve high luminosity in a colliding beam accelerator, the amplitude function is made small at the point where the beams are brought into collision. The length of the detector occupying this straight section will be large compared to the value of β at the interaction point. Show that the phase advance through this straight section will be approximately 180°.

18. Show that $J^2 = -I$. Show that n repetitions of

$$M = I \cos \mu + J \sin \mu$$

give the result akin to de Moivre's theorem,

$$M^n = I \cos n\mu + J \sin n\mu.$$

19. The discussion in this chapter has been explicitly carried out in the context of stable motion, where amplitude functions and phase advances

are real numbers. Of course, the formalism works equally well for unstable motion. Consider the following case. Suppose one starts with a synchrotron with a one-turn matrix given by

$$
M_0 = \begin{pmatrix} 0 & \beta_0 \\ -\dfrac{1}{\beta_0} & 0 \end{pmatrix}
$$

and adds a thin lens quadrupole of focal length $f = 4\beta_0$. Evaluate the new amplitude function at this point and the new phase advance modulo 2π.

20. Suppose that a 10 GeV/c proton beam with normalized (39%) emittance of 2π mm mrad is injected into a synchrotron having a half cell length of 30 m and a cell phase advance of 68°. Estimate the boundaries of the beam excursions in displacement and angle.

21. The synchrotron into which the beam of the preceding problem is to be injected offers a half aperture of 50 mm in the horizontal plane. Calculate the admittance. Normally it is necessary that the admittance be much larger than the beam emittance for reasonable performance.

22. Suppose that there are many uncorrelated angular deflections distributed around a ring which average to zero and have an rms value θ_{rms}.
    (a) Show that the rms (over an ensemble of synchrotrons) orbit distortion at some point of observation where the amplitude function is β_0 is

$$
\langle x^2 \rangle^{1/2} = \frac{\beta_0^{1/2}}{2|\sin \pi \nu|} \left(\sum_i \theta_{rms}^2 \beta_i \sin^2(\pi \nu - \psi_i) \right)^{1/2}
$$

$$
= \frac{(\beta_0 \bar{\beta})^{1/2}}{2\sqrt{2}\,|\sin \pi \nu|} N^{1/2}\theta_{rms},
$$

where $\bar{\beta}$ is the average of the amplitude function at the N kick locations. In proceding from the first to the second form of the result, note that a term of order unity compared with N is neglected; for many purposes, this is an excellent approximation.
    (b) Calculate the rms orbit distortion expected in a 20 TeV scale synchrotron, if the only source of the angular deviations is quadrupole alignment errors characterized by an rms value of 1 mm. For amplitude functions and focal lengths, use values characteristic of a thin lens FODO cell with a length of 180 m and a phase advance of 90°.

23. In Problem 22, only the rms orbit distortion was calculated. One would like to know the distribution of peak distortions, or, to put it more

precisely, to know the odds of the maximum distortion exceeding the rms by a given factor. This is an easy job for a small computer. Generate an ensemble of synchrotrons, each populated with a set of steering errors having unit standard deviation and following a Gaussian distribution. Find the maximum orbit deviation for each synchrotron in units of the rms and plot the integral distribution. You should find, for example, that the probability of the peak distortion exceeding the rms value by a factor of two is 60%.

24. Orbits can be corrected and adjusted by using steering dipoles. One standard algorithm is based on so-called *three-bumps*. A local orbit distortion can be made by three steering dipoles. Let the three steering angles be θ_1, θ_2, and θ_3. Show that if these angles are related according to

$$\theta_2 = -\theta_1 \left(\frac{\beta_1}{\beta_2}\right)^{1/2} \frac{\sin \psi_{13}}{\sin \psi_{23}},$$

$$\theta_3 = \theta_1 \left(\frac{\beta_1}{\beta_3}\right)^{1/2} \frac{\sin \psi_{12}}{\sin \psi_{23}}$$

then the orbit distortion is localized between the first and third steering elements.

25. If a single quadrupole is hooked up backwards, so its focusing character is the reverse of what it should be, is it likely that the betatron oscillations will still be stable in both transverse planes? As usual, use a ring of simple FODO cells for this problem.

26. In the same spirit as Problem 22, derive an expression for the rms tune shift arising from uncorrelated gradient errors. Apply your expression to the case of a 20 TeV scale ring in which the main quadrupoles exhibit an rms fractional error in their focal lengths of 0.1%.

27. Suppose that a quadrupole of negligible length l and strength $q \equiv B'l/(B\rho)$ is placed in a ring at a point where the amplitude function has value β_1. Assume that upstream of this point the amplitude function is unperturbed. Show that downstream of the quadrupole the fractional deviation in β is given by

$$\frac{\Delta \beta}{\beta} \equiv \frac{\beta(s) - \beta_0(s)}{\beta_0(s)} = -q\beta_1 \sin 2\psi_0 + \frac{1}{2}(q\beta_1)^2(1 - \cos 2\psi_0)$$

where $\beta_0(s)$ is the original amplitude function, and the original phase ψ_0 is measured from the location of the quadrupole.

28. For sufficiently small quadrupole errors, the nonlinear term in the equation of the preceding problem can be neglected. Then the fractional

deviation in the amplitude function obeys rules identical to the orbit distortions which arise from steering errors. Show that the rms fractional change in β associated with quadrupole errors is, in this approximation,

$$\left(\frac{\Delta\beta}{\beta}\right)_{rms} = \frac{1}{2\sqrt{2}\,|\sin 2\pi\nu|}\left\langle\sum_i q_i^2\beta_i^2\right\rangle^{1/2},$$

where the q's are defined as in the preceding problem.

29. Let the bending of a thin lens FODO cell be performed by pure dipole magnets providing a bend angle θ with the bend center at the midpoint of each half cell. Show that the maximum and minimum values of the dispersion function for a ring made entirely of such cells are

$$D_{max,\,min} = \frac{\theta L}{\sin^2 \frac{1}{2}\mu}\left(1 \pm \tfrac{1}{2}\sin\tfrac{1}{2}\mu\right),$$

where L is the half-cell length. For typical proton ring designs, the product θL tends to be about 1 meter. Since the phase advance per cell, μ, also is apt to be selected from a narrow range, the dispersion is about the same regardless of the peak energy of the synchrotron. It will be interesting to see if this scaling perseveres in the future.

30. Suppose that the basic repeat period of a synchrotron consists of $n-1$ FODO cells of the sort used in the preceding problem, followed by a single FODO cell without bending. Show that in the bending cells the total dispersion is the sum of that appropriate to the FODO cell alone and a free oscillation having initial values (at the beginning of the superperiod)

$$\left(\begin{matrix}\Delta D\\ \Delta D'\end{matrix}\right) = \frac{(I - M^{-n})(M - I)}{2(1 - \cos n\mu)}\left(\begin{matrix}D\\ D'\end{matrix}\right)_{\text{FODO only}},$$

where M is the 2×2 matrix for propagation of a betatron oscillation through each cell. Note that the dispersion can now take on arbitrarily large positive or negative values. This is an example of a mismatched lattice function; the juxtaposition of cells with different intrinsic dispersion functions can lead to a possible unacceptable total dispersion.

31. Extend the phase-amplitude form of the transfer matrix to include momentum dispersion. The three additional elements needed are

$$m_{13} = D_2 - m_{11}D_1 - m_{12}D_1',$$
$$m_{23} = -m_{21}D_1 - m_{22}D_1' + D_2',$$
$$m_{13} = 1,$$

where the m's on the right hand side are elements of the 2×2 matrix as written down in Section 3.2 for transport of a betatron oscillation from point 1 to point 2.

32. To illustrate the principle of matching of lattice functions, here is one method of proceeding from a bend (nonzero dispersion) region to a straight section without setting up a dispersion wave. Coming from the arc, the beam first encounters a FODO cell in which the bending is reduced to a fraction $1 - x$ of the standard cell deflection and then a cell in which the bending fraction is x. As before, all bend centers are at the midpoint of the half cell. Prove that the condition for the dispersion function and its derivative to be unchanged in the arc and vanish at the entry to the straight section is

$$ x = \frac{1}{2(1 - \cos \mu)}, $$

where μ is the phase advance of each of the FODO cells making up the ring. It may be useful to use the matrix of the preceding problem, and decide what happens to m_{13} and m_{23} as the bend strength is changed.

33. For the simple FODO lattice synchrotron, verify that the transition gamma, γ_t, is about equal in magnitude to the tune.

34. A steering error of strength θ produces a new closed orbit in a synchrotron which may have a path length different from the ideal path length by an amount ΔC. Show that $\Delta C = \theta D$, where D is the value of the dispersion function at the location of the steering error.

35. For the simple FODO lattice synchrotron, verify that the chromaticity is about equal in magnitude and opposite in sign to the tune if the only elements contributing to the chromaticity are the main quadrupoles.

36. Assume that the bending magnets exhibit a systematic sextupole moment. The field can be written in the form

$$ B = B_0\left(1 + b_2 x^2\right), $$

where $b_2 = B''/2B_0$ is the sextupole coefficient. Show that the associated contribution to the chromaticity is

$$ \Delta \xi = \pm \langle D\beta b_2 \rangle $$

with the sign depending on the degree of freedom. Under some circumstances, $\Delta \xi$ can be large. At low fields, superconducting magnets can have signicant sextupole moments due to persistent currents. A possible value for b_2 is 3 m^{-2}. Estimate the resulting chromaticity for a 20 TeV scale ring.

37. Derive the expressions for chromaticity adjustment written down in the last section.

38. Consider a *dogleg* as shown in the diagram below.

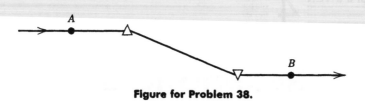

Figure for Problem 38.

(a) If the dispersion function at A is $D = D' = 0$, what is the value of the dispersion function at B? (Assume that $\theta \approx d/L \ll 1$.)

(b) At Fermilab, the 150 GeV Main Ring injector is located 25.5 in. above the Tevatron synchrotron. Beams are transferred between the two rings in a long straight section, free of quadrupoles. How much vertical dispersion is "generated" just by the transfer process?

(c) Since the design vertical dispersion function of the Tevatron is zero everywhere, there would be a mismatch of the dispersion function between the Main Ring and Tevatron if left uncorrected. Would you expect this mismatch to affect the emittance of the beam after injection? Why or why not?

CHAPTER 4

Resonances and Transverse Nonlinear Motion

Despite our attempt to concentrate on linear behavior in the last chapter, we found ourselves compelled at the end to introduce nonlinear magnetic fields to compensate chromatic aberration. In fact, as accelerators have grown in energy, cost, and performance demands, it has become necessary to devote more and more attention to the effects of nonlinear fields on the single particle dynamics of synchrotrons, both those necessitated by the design and those arising from magnet imperfections.

In the first category are the present generation of synchrotron radiation sources. In these devices the emphasis on small emittance to produce high brightness beams results in very strong sextupole elements to compensate chromaticity. The attendant aberrations lead to a bound on the stable region in transverse phase space. The stable region is often called the *dynamic aperture*.

In the second category are large hadron colliders. Here, one is playing cost against the provision of the design magnetic field. Again, the presence of nonlinearities will lead to a finite dynamic aperture, and the designer must assure that performance goals are met without crossing the border into overdesign.

The introduction of a single nonlinear element can change dramatically the mappings in phase space, which in the last chapter were simple ellipses. The nonlinear equations of motion have the disadvantage that they cannot be solved in closed form. However, the iteration of the associated difference equations can demonstrate many of the essential features; all one needs is a home computer. For example, suppose there is a single thin sextupole installed in a synchrotron which has otherwise perfectly linear fields. For a

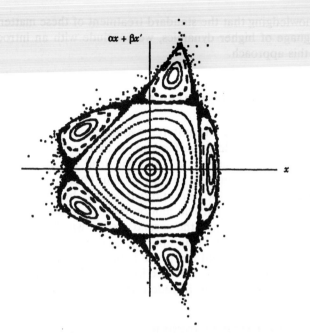

Figure 4.1. Phase space plot of particles circulating in a synchrotron lattice that is perfectly linear except for a single thin sextupole field introduced at the point of observation. For this plot, the tune of the accelerator is 0.42 plus an integer.

specific choice of parameters, the graphics output from a turn-by-turn iteration of the equations of motion is shown in Figure 4.1.

In the purely linear case, particle trajectories in this phase space lie on circles; for sufficiently small amplitudes this is still the case, as seen in the figure. However, as one moves to larger amplitudes, the nonlinearity manifests itself and the trajectories increasingly deviate from circles. At still larger radius, we encounter a set of five *islands*, isolated regions of stability. At slightly larger amplitude the motion becomes irregular, or *chaotic*. Finally, at large enough initial amplitudes the motion is completely unstable. If there were a clearly defined separatrix, as there was in the case of longitudinal dynamics, the definition of the dynamic aperture would be clear. But already, in this simple numerical example, we see that the richer phase space dynamics revealed here makes a straightforward definition of dynamic aperture difficult.

In this text we will not attempt to unravel all of the features of Figure 4.1; more extensive discussions may be found in the advanced references in the bibliography. Our purpose here will be to examine simple situations in which the equations of nonlinear motion are tractable. We will extend the discussion of resonances to include those of nonlinear origin, and interestingly enough these cases rather directly permit progress toward the solution of the equations of motion.

Acknowledging that the standard treatment of these matters is couched in the language of higher dynamics, we conclude with an introductory exposition of this approach.

4.1 TRANSVERSE RESONANCES

In the last chapter, we encountered two instances of resonant behavior, both stated in terms of field imperfections. If there are steering errors, however small, an integer value of the tune will lead to an oscillation growing in amplitude linearly from turn to turn without bound. The "without bound" is of course an artifact of our approximations, but because of the limited aperture in which the particle motion must be accommodated, the approximations are reasonable. In the second case, it was found that at a tune such that $2\nu = m$ where m is an integer, any quadrupole error will lead to amplitude growth. We did not prove it then, but the growth in this case can be faster than linear. In this section, we will attempt to generalize the discussion of resonances to include nonlinear field perturbations.

4.1.1 Floquet Transformation

We begin by completing the coordinate transformation so that the resulting motion is indeed just that of a simple harmonic oscillator. Start from the solution to the equation of motion:

$$x(s) = A\beta^{1/2}(s)\cos[\psi(s) + \delta]. \qquad (4.1)$$

If we define a "reduced phase," ϕ, by $\phi = \psi/\nu$, then ϕ is a variable that increases by 2π for each turn. Even though ϕ is not a real polar angle measured from the center of a circle, it behaves like one. If now one defines a new dependent variable ζ according to

$$\zeta = \frac{x}{\beta^{1/2}}, \qquad (4.2)$$

then

$$\zeta(\phi) = A\cos(\nu\phi + \delta) \qquad (4.3)$$

and the free betatron oscillation reduces to simple harmonic motion, with ν oscillations for every advance of ϕ by 2π. The equation of motion for ζ is just

$$\frac{d^2\zeta}{d\phi^2} + \nu^2\zeta = 0. \qquad (4.4)$$

The replacement of coordinates x, s by ζ, ϕ is called a Floquet transformation. A direct benefit of the Floquet transformation can be seen as follows. The equation of motion was developed for an accelerator without field errors. With field errors, one of the problems at the end of the chapter is to show that the right hand side of the above equation will no longer be zero; we will have instead

$$\frac{d^2\zeta}{d\phi^2} + \nu^2\zeta = -\nu^2\beta^{3/2}\frac{\Delta B(\zeta, \phi)}{(B\rho)}, \qquad (4.5)$$

where ΔB represents all those fields not taken into account in setting up the design orbit.

Therefore, in coordinates ζ, ϕ, the full collection of mathematical methods for treating driven harmonic oscillations becomes available, and the notion of a resonance between some harmonic amplitude of the right hand driving term and the tune ν is just the same as in the case of a simple oscillator.

4.1.2 Multipole Expansion

The next task is to choose a method in which to express the field error ΔB. For many purposes, it is convenient to use a multipole expansion to go beyond the at most linear dependence on x to which we have limited ourselves thus far. But rather than take up the general case (which is left to Chapter 5), let us stay for the present in one degree of freedom and take

$$\Delta B = B_0(b_0 + b_1 x + b_2 x^2 + \cdots), \qquad (4.6)$$

where B_0 is a reference field strength and the b_n are the multipole coefficients. In a bending magnet, for instance, B_0 would be the nominal bend field. For definiteness, we are taking the case of motion in the bending plane; ΔB is the variation in the vertical component of the field on the midplane, and, in "pole language" the b_n arise from normal multipole errors. That is to say, the field imperfections arise from pole distributions that do not have poles in the horizontal plane.

So b_0 is the dipole error, b_1 is the quadrupole error term, b_2 the sextupole term, and so on. The b_n are of course functions of s.

Some examples of how these errors arise might be useful. Suppose that all the bending magnets are intended to have a field of exactly 1 tesla, at a certain excitation current common to all the magnets. For a bending magnet such as that sketched in Figure 4.2, the field in the gap in the infinite permeability approximation is

$$B = \mu_0 NI/g, \qquad (4.7)$$

where NI is the number of ampere-turns, g is the gap height, and the field

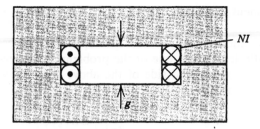

Figure 4.2. Dipole bending magnet with gap height g and N turns of conductor, each carrying current I.

will be a constant within the coil window. The laminations of which the steel yoke are typically made are produced by stamping them out of steel sheets. Such factors as wear of the die used in the stamping, or the use of multiple dies for a large production run, will lead to a variation in the gap height. If g is designed to be exactly 5 cm, then a change of 0.0025 cm (approximately one-thousandth of an inch) will give $b_0 = 5 \times 10^{-4}$, which is a significant field error.

Such magnets are often assembled from two "top and bottom" half cores, as suggested by the horizontal lines at the midplane in the sketch. If the cores meet perfectly on one side, but are separated by a small gap h on the other, it is easy to show that the resulting quadrupole term will be

$$ b_1 = \frac{h}{gw}, \tag{4.8} $$

where w is the pole width. For $h = 0.0025$ cm in the same magnet, and $w = 10$ cm, then $b_1 = 0.5 \times 10^{-4}$/cm, again a significant error.

The most pernicious source of sextupole terms in both conventional and superconducting magnets is remanent magnetization, arising from finite remanence in ferromagnets and persistent currents in superconducting magnets. Rather than digress into a discussion of material properties, let us rather illustrate the appearance of sextupole terms in a simpler but equally common situation—that due to eddy currents in a conducting vacuum chamber.

In Figure 4.3, we have added a vacuum chamber to a dipole magnet. If the magnetic field is changing with time, according to Faraday's law there will be an emf induced in a loop characterized by positions $\pm x$ from the chamber center as illustrated in the right hand portion of the figure. The electric field is

$$ E = \dot{B}x, \tag{4.9} $$

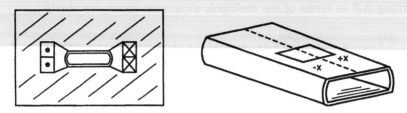

Figure 4.3. Dipole magnet with steel vacuum chamber.

and if the material has conductivity σ, the current density is

$$j = \sigma E = \sigma \dot{B} x. \tag{4.10}$$

The total current within the area between $x = 0$ and x is then

$$I = h\sigma \dot{B} x^2, \tag{4.11}$$

where h is the thickness of the chamber, and the contributions of both top and bottom have been included. Therefore, the field difference between the center of the magnet and x is

$$\Delta B = \mu_0 I/g = \mu_0 h\sigma \dot{B} x^2/g \tag{4.12}$$

which is a sextupole field, with moment

$$b_2 = \mu_0 \sigma \frac{h}{g} \frac{\dot{B}}{B}. \tag{4.13}$$

As an example, let's consider the Main Ring synchrotron at Fermilab. The vacuum chamber is stainless steel for which $\mu_0 \sigma \approx 2 \ \text{sec}/\text{m}^2$. For h/g, take $\frac{1}{30}$, and for $\dot{B}/B$ use $10/\text{sec}$. Then $b_2 \approx \frac{2}{3} \ \text{m}^{-2}$. Again, this is a significant field perturbation. To verify that the aforementioned field errors are indeed significant, the expressions for closed orbit distortions, tune shift, and chromaticity derived in the last chapter can be employed, using the table of Fermilab Main Ring parameters found in the Appendices.

4.1.3 The Driven Oscillator and Rational Numbers

We may now return to the equation of motion for a betatron oscillation as written after the Floquet transformation, Equation 4.5:

$$\frac{d^2\zeta}{d\phi^2} + \nu^2\zeta = -\nu^2\beta^{3/2}\frac{\Delta B(\zeta,\phi)}{(B\rho)}. \tag{4.14}$$

Writing ΔB in terms of the multipole expansion above, we obtain

$$\frac{d^2\zeta}{d\phi^2} + \nu^2\zeta = -\frac{\nu^2 B_0}{(B\rho)}\left[(\beta^{3/2}b_0) + (\beta^{4/2}b_1)\zeta + (\beta^{5/2}b_2)\zeta^2 + \cdots\right].$$

$$(4.15)$$

A term on the right-hand side with the same frequency as the natural frequency, ν, of the oscillator would be cause for concern. The products of amplitude functions and multipole coefficients can be expressed as Fourier series in ϕ:

$$(\beta^{(n+3)/2}b_n) = \sum_k c_k e^{\pm ik\phi},$$

$$(4.16)$$

while the solution to the inhomogeneous equation of motion can be written as

$$\zeta(\phi) = \zeta_0 e^{\pm i\nu\phi}.$$

$$(4.17)$$

Consider the first term; b_0 represents the dipole field error. If the product $\beta^{3/2}b_0$ has a nonvanishing kth harmonic at $k = \nu$, a resonant condition will exist. We already know that integral values of the tune must be avoided.

The next term contains the gradient errors, characterized by b_1. The kth harmonic of the factor $\beta^2 b_1$ can beat with the frequency ν presented by ζ to produce a driving frequency $k - \nu$. The resonance condition $k - \nu = \nu$ gives $k = 2\nu$; i.e., the tune shouldn't be a half integer. Again, we already know that. (The beat frequency with the plus sign, $k + \nu = \nu$, is a special case; the zeroth harmonic of $\beta^2 b_1$ is a tune shifting term and represents a "renormalization" of the left hand side of the equation of motion rather than a resonance.)

The third term represents the effect of sextupole moments. The factor ζ^2 can exhibit a frequency 2ν; when this is combined with the kth harmonic of $\beta^{5/2}b_2$, one can have the condition $k - 2\nu = \nu$, or $k = 3\nu$. That is, the tune should not be a third of an integer. The beat frequency $k + 2\nu$ just leads again to the condition that the tune should not be an integer.

In general, any tune of the form $\nu = k/n$ can resonate with some multipole moment. The integer in the denominator is called the order of the resonance; for instance, sextupoles can produce third order resonances, octupoles can produce fourth order resonances, and so on.

This argument can only suggest that problems can arise for tunes equal to some rational numbers. Both experience and more quantitative arguments indicate that low order resonances need to be avoided. How low is low depends on the application and resonance strength.

A distinction is made between driving terms that arise from random field errors and those that arise from systematic imperfections common to all of the magnets. The language used in the discussion above was in the spirit of the resonances having their origin in random field errors—construction tolerances result in each magnet being slightly different, and so all harmonics are represented in the resonance driving terms. The magnets may also possess systematic nonlinear multipoles; the overall symmetry of the ring will play a role in the presence or absence of particular harmonics. Suppose the overall lattice consists of P identical periods; the only harmonics arising from systematic field imperfections will be of the form Pk, $k = 1, 2, 3, \ldots$, and the resonant tunes will be Pk/n. The greater the symmetry of the ring, the easier it is to stay away from systematic resonances. For example, if a ring is made up of six identical superperiods and the bending magnets have a systematic sextupole field error, the systematic third order resonant tunes are the even integers. Highly symmetric synchrotrons are generally rather small. In very large accelerators, cost and operational considerations militate against the preservation of high symmetry. Therefore, corrector systems must under-take the role of compensation of unavoidable systematic field errors.

Even though we've worked in only one degree of freedom, we should at least quote the result for the complete transverse case. The resonances are lines in the ν_x, ν_y plane of the form

$$M\nu_x + N\nu_y = P, \tag{4.18}$$

where M, N, and P are integers all of the same sign (one of the pair M, N can be zero) for instability. Justification of this result will be provided in the next chapter. The sum of the absolute values of M and N is the order of the resonance, and the order can be related to a multipole term just as in the one degree of freedom case. If M and N are opposite in sign, the result is coupled but stable motion. Take the sextupole case again. There are four sum (hence, unstable) resonances:

$$3\nu_x = P, \tag{4.19}$$

$$2\nu_x + \nu_y = P, \tag{4.20}$$

$$\nu_x + 2\nu_y = P, \tag{4.21}$$

$$3\nu_y = P. \tag{4.22}$$

The first and third are driven by the sextupole term in the field expansion used above. The second and fourth are driven by a sextupole field, but one rotated by 30° with respect to the first to form a *skew* sextupole.

4.2 A THIRD-INTEGER RESONANCE

Having introduced the notion of nonlinear resonances in the previous sec-tion, the point now is to try to turn the purely qualitative approach into something that can claim to be a quantitative treatment. Let's take the case of a sextupole nonlinearity distributed in an arbitrary fashion around a circular accelerator. There are several approaches one may take to looking at the problem, including Fourier analysis, Hamiltonian perturbation theory, and computer simulation. For this discussion, we will treat the nonlinearity as a small perturbation of the linear motion. We will assume that the linear tune is very close to one-third of an integer, so that we may expect that the perturbation of the linear motion will be dominated by the sextupole field. On each turn, we add up the effects of the nonlinearities as though they were independent of each other, and take stock of the situation at one point on the ring after each turn.

4.2.1 Equation of Motion

We may write the linear oscillation as

$$x = a\left(\frac{\beta(s)}{\beta_0}\right)^{1/2} \cos \chi(s), \qquad (4.23)$$

where the amplitude function, β_0, at the point of observation has been used explicitly so that the symbol a can denote a real amplitude at that point. The symbol for the phase has been switched to χ so that ψ can be reserved for the phase at the point of observation.

For the other coordinate of phase space, we use

$$p_x \equiv \beta(s)x' + \alpha(s)x = -a\left(\frac{\beta(s)}{\beta_0}\right)^{1/2} \sin \chi, \qquad (4.24)$$

and the unperturbed motion at any point in the ring is just a circle (see Figure 4.4). (Note that p_x is not the transverse kinematic momentum.)

Assume a magnetic field $\Delta B(x, s)$, perpendicular to x and s, is introduced at s and extends over a length Δs. The sign of ΔB is positive if ΔB is

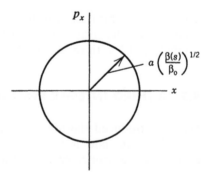

Figure 4.4. Phase space coordinates.

directed in the same sense as the main guide field. For sufficiently small Δs, x does not change as the particle traverses Δs, but the slope does, according to

$$\Delta x' = -\frac{\Delta B\,\Delta s}{(B\rho)}, \tag{4.25}$$

and so

$$\Delta p_x = -\beta(s)\frac{\Delta B\,\Delta s}{(B\rho)}. \tag{4.26}$$

As a result of the perturbation, the amplitude and phase have changed. From

$$\Delta x = \left(\frac{\beta}{\beta_0}\right)^{1/2}(\Delta a\cos\chi - a\sin\chi\cdot\Delta\chi) = 0, \tag{4.27}$$

$$\Delta p_x = -\left(\frac{\beta}{\beta_0}\right)^{1/2}(\Delta a\sin\chi + a\cos\chi\cdot\Delta\chi) = -\beta\frac{\Delta B\,\Delta s}{(B\rho)}, \tag{4.28}$$

one finds

$$\Delta a = \frac{\beta_0}{(B\rho)}\left(\frac{\beta}{\beta_0}\right)^{1/2}\Delta B\,\Delta s\,\sin\chi, \tag{4.29}$$

$$\Delta\chi = \frac{\beta_0}{(B\rho)}\left(\frac{\beta}{\beta_0}\right)^{1/2}\frac{\Delta B\,\Delta s}{a}\cos\chi. \tag{4.30}$$

Now we add up these perturbations over a turn. Suppose that at a given passage of our point of observation, the phase of the oscillation is ψ. Then on the succeeding turn, the unperturbed phase χ would develop according to

$$\chi(s) = \psi + \nu\phi(s), \tag{4.31}$$

where

$$\phi(s) \equiv \int\frac{ds}{\nu\beta(s)}. \tag{4.32}$$

To obtain the first order equations of motion, we assume that the changes in amplitude and phase can be found by adding up the individual perturbations.

For the amplitude we immediately obtain from Equation 4.29

$$\frac{da}{dn} = \frac{\beta_0}{(B\rho)} \oint \left(\frac{\beta}{\beta_0}\right)^{1/2} \Delta B(x,s) \sin[\psi + \nu\phi(s)] \, ds. \tag{4.33}$$

The change in phase at the observation point after passing through one turn is $2\pi\nu$ plus the accumulated phase change due to the sum of all the contributions typified by Equation 4.30:

$$\Delta\psi = 2\pi\nu + \frac{\beta_0}{(B\rho)} \oint \left(\frac{\beta}{\beta_0}\right)^{1/2} \frac{\Delta B(x,s)}{a} \cos[\psi + \nu\phi(s)] \, ds. \tag{4.34}$$

It is the difference between $\Delta\psi$ and $2\pi\nu$ that is small in the spirit of our perturbation calculation. Therefore, the differential equation for phase advance will be

$$\frac{d}{dn}(\psi - 2\pi\nu n) = \frac{\beta_0}{(B\rho)} \oint \left(\frac{\beta}{\beta_0}\right)^{1/2} \frac{\Delta B(x,s)}{a} \cos[\psi + \nu\phi(s)] \, ds. \tag{4.35}$$

4.2.2 Recognition of the Sextupole Resonance

So far nothing has been said about the variety of the nonlinearity represented by $\Delta B(x,s)$. Now we take the case of a sextupole distribution:

$$\Delta B(x,s) = \frac{B''(s)}{2} x^2. \tag{4.36}$$

Insertion of this form of the field into the equation of motion for the amplitude as well as Equation 4.23 yields, after some manipulation of the trigonometric functions,

$$\frac{da}{dn} = \tfrac{1}{4}a^2 \frac{\beta_0}{(B\rho)} \oint \left(\frac{\beta}{\beta_0}\right)^{3/2} \left(\frac{B''}{2}\right)$$

$$\times (\sin\psi \cos\nu\phi + \cos\psi \sin\nu\phi + \sin 3\psi \cos 3\nu\phi + \cos 3\psi \sin 3\nu\phi) \, ds. \tag{4.37}$$

We now look for terms that could be additive from turn to turn, that is, terms that could represent unstable motion. If the tune were close to an integer, the first two terms could be candidates. But if the tune is not near an integer (and we assume that it is not), $\sin\psi$ and $\cos\psi$ will change rapidly from turn to turn, and so the amplitude will not grow steadily. However, if 3ν were an integer, $\sin 3\psi$ and $\cos 3\psi$ would have constant values from turn to turn, and then the amplitude could exhibit growth. So we ignore the first two terms and retain the second pair.

Since we want to study the case where 3ν is not exactly an integer, but close, let $3\nu_0$ denote the integer of interest, with the tune difference, $\delta \equiv \nu - \nu_0$, small compared with unity. The equation of motion for the amplitude then can be written

$$\frac{da}{dn} = \tfrac{1}{4}a^2(A\sin 3\psi + B\cos 3\psi) \tag{4.38}$$

with

$$A \equiv \frac{\beta_0}{(B\rho)} \oint \left(\frac{\beta}{\beta_0}\right)^{3/2} \left(\frac{B''}{2}\right)\cos 3\nu_0\phi \, ds, \tag{4.39}$$

$$B \equiv \frac{\beta_0}{(B\rho)} \oint \left(\frac{\beta}{\beta_0}\right)^{3/2} \left(\frac{B''}{2}\right)\sin 3\nu_0\phi \, ds. \tag{4.40}$$

In defining A and B, we have used the proximity of ν to ν_0 so that A and B are true harmonic amplitudes.

The equation of motion for ψ is found by the same procedure, but with one modification. As a result of the unperturbed motion alone, the phase advances by $2\pi\nu$ in one turn, and so ψ by itself hardly qualifies as a continuous variable. This circumstance was already recognized in writing the left hand side of Equation 4.35. Now observe that the factors related to ψ that enter the right hand sides of the equations, $\cos 3\psi$ and $\sin 3\psi$, are insensitive to the replacement of ψ by $\psi - 2\pi\nu_0 n$. We introduce a new variable $\tilde{\psi} \equiv \psi - 2\pi\nu_0 n$, and obtain the equations of motion in terms of the new phase:

$$\frac{da}{dn} = \tfrac{1}{4}a^2\left(A\sin 3\tilde{\psi} + B\cos 3\tilde{\psi}\right), \tag{4.41}$$

$$\frac{d\tilde{\psi}}{dn} = \tfrac{1}{4}a\left(A\cos 3\tilde{\psi} - B\sin 3\tilde{\psi}\right) + 2\pi\delta. \tag{4.42}$$

With the foregoing redefinition, $\tilde{\psi}$ is not only a variable continuous in n, but the form of Equation 4.38 is preserved with the replacement of ψ by $\tilde{\psi}$. We have, in effect, made a transformation to rotating coordinates.

4.2.3 First Integral and the Separatrix

The equations of motion were developed in the phase-amplitude form because the characteristic of an unstable resonance is more readily identified there. The transformation back to (rotating) Cartesian coordinates $\tilde{x}, \tilde{p}_x$

follows from

$$\frac{d\tilde{x}}{dn} = \frac{\tilde{x}}{a}\left(\frac{da}{dn}\right) + \tilde{p}_x\left(\frac{d\tilde{\psi}}{dn}\right), \tag{4.43}$$

$$\frac{d\tilde{p}_x}{dn} = \frac{\tilde{p}_x}{a}\left(\frac{da}{dn}\right) - \tilde{x}\left(\frac{d\tilde{\psi}}{dn}\right), \tag{4.44}$$

and gives

$$\frac{d\tilde{x}}{dn} = \tfrac{1}{4}A(-2\tilde{x}\tilde{p}_x) + \tfrac{1}{4}B(\tilde{x}^2 - \tilde{p}_x^2) + 2\pi\delta \cdot \tilde{p}_x, \tag{4.45}$$

$$\frac{d\tilde{p}_x}{dn} = \tfrac{1}{4}A(\tilde{p}_x^2 - \tilde{x}^2) + \tfrac{1}{4}B(-2\tilde{x}\tilde{p}_x) - 2\pi\delta \cdot \tilde{x}. \tag{4.46}$$

For simplicity consider the case $B = 0$. A first integral of this system is

$$\left(\tilde{x} - \frac{4\pi\delta}{A}\right)\left[\tilde{p}_x^2 - \tfrac{1}{3}\left(\tilde{x} + \frac{8\pi\delta}{A}\right)^2\right] = \text{constant} \equiv k, \tag{4.47}$$

and different phase space trajectories are associated with different values of k. This first integral can be found by the usual mixture of technique and guesswork associated with solving differential equations. Or one might argue that the equations of motion in $\tilde{x}$ and $\tilde{p}_x$ are some version of Hamilton's equations, and so there ought to be a function $\mathscr{H}$ such that

$$\frac{d\tilde{x}}{dn} = \frac{\partial\mathscr{H}}{\partial\tilde{p}_x}, \tag{4.48}$$

$$\frac{d\tilde{p}_x}{dn} = -\frac{\partial\mathscr{H}}{\partial\tilde{x}}; \tag{4.49}$$

and the first integral can indeed by found by pursuing this course.

Notice that there are fixed points which satisfy $d\tilde{x}/dn = d\tilde{p}_x/dn = 0$ and are located at the points

$$\left(\tilde{x} = -\frac{8\pi\delta}{A}, \tilde{p}_x = 0\right), \tag{4.50}$$

and

$$\left(\tilde{x} = \frac{4\pi\delta}{A}, \tilde{p}_x = \pm\sqrt{3}\,\frac{4\pi\delta}{A}\right). \tag{4.51}$$

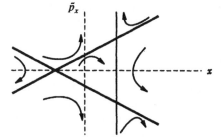

Figure 4.5. Separatrix when near the third-integer resonance. The arrows indicate direction of flow of phase space trajectories.

For these $\tilde{x}, \tilde{p}_x$, the constant k is zero and the associated figure in phase space—the separatrix—is just three intersecting straight lines. Motion within the triangle is bounded; motion outside the triangle is unbounded in this approximation (see Figure 4.5).

Along the vertical separatrix, the equation of motion becomes

$$\frac{d\tilde{p}_x}{dn} = \tfrac{1}{4}A\tilde{p}_x^2 - \tfrac{3}{4}A\left(\frac{4\pi\delta}{A}\right)^2, \tag{4.52}$$

which is easily integrated to yield $\tilde{p}_x$ as a function of n. (See the problems at the end of the chapter.) Motion along or near the separatrix is important in resonant beam extraction as will be seen in the next subsection.

Finally, we can define the width of a nonlinear resonance, or at least present one of several similar definitions. There is a qualitative difference between the linear resonance produced, for example, by gradient errors and a nonlinear resonance. In the former case, the entire beam is either stable or unstable. In the latter case, the motion may be stable or unstable, depending on the oscillation amplitude. A nonlinear resonance doesn't produce a stopband—an essentially linear notion—but a range of tunes of the linear lattice throughout which varying fractions of the beam are unstable.

Suppose the beam has emittance $\epsilon = \pi\sigma^2/\beta_0$. If the tune is initially far from resonance and the resonance is approached sufficiently slowly, then it is reasonable to suppose (since the supposition can be easily checked by a simulation) that the phase space area will gradually deform from its originally circular boundary into the triangular shape characteristic of the resonance. When the tune difference is such that the stable area is equal to the beam emittance, one can say with some justification that 2δ is a reasonable definition of the width of the resonance. For the case that we have been considering, the resonance width is given by

$$2\delta = \frac{A}{2\pi}\left(\frac{2\beta_0\epsilon}{\sqrt{3}}\right)^{1/2}; \tag{4.53}$$

here the emittance ϵ contains 39% of the particles. Obviously, there is some arbitrariness in such a definition. In the process of injection into a ring, the sudden rather than the adiabatic approximation would be reasonable.

4.2.4 Application to Resonant Extraction

Much of the foregoing discussion emphasized undesirable consequences of nonlinearities. The controlled introduction of nonlinearities may be used to advantage, however, and one important instance is the process of resonant extraction. It is easy to take the particles in an accelerator out in one turn. Unfortunately, that approach to extraction does not necessarily satisfy the needs of the experimental program. Rather, it is usually the case that particles are to be dribbled out on a time scale of a millisecond to many seconds. So we resort to the resonant extraction process, the groundwork for which has been established in the previous subsection.

In this process, the separatrix is made to gradually squeeze the phase space occupied by the beam. No matter how slowly this "squeeze" is carried out, motion near the unstable fixed points will fail to be adiabatic; for in the neighborhood of these points motion becomes arbitrarily slow in the static case. Particles depart the stable area at the fixed points and stream out along the outgoing arms of the separatrix. But the continuum in amplitudes of these outgoing particles leads to beam loss, since somewhere there has to be a device that is the start of the channel for departing particles. The partition, or septum, between "in" and "out" particles must be as thin as possible in order to obtain high extraction efficiency. At present, the thin septa are electrostatic. They are not strong enough to fling particles out of the ring, but they can establish enough of a gap between "in" and "out" particles so that a second, stronger magnetic septum can direct particles into an exit channel. The great virtue of the electrostatic septum is that the obstacle presented to the beam is very small, about 50 μm. For a proton accelerator with conventional magnets, it is already important that the losses off the primary extraction septum be minimized: the buildup of radioactivity in the ring is a major concern for operation and maintenance. For a proton accelerator using superconducting magnets, control of the septum loss is essential to avoid excessive energy deposition in the magnets with the attendant risk of superconducting to normal transition; that is, the extraction process brings with it the danger of a quench.

The third integer resonance can be used for slow extraction. The phase space at the first septum might look like the sketch in Figure 4.6(a), where the coordinates are the x and p_x in the unrotating frame. The septum is located a distance x_s from the center of the aperture. The *step size*, Δ, is the growth in x in three turns. The orientation of the figure is determined by the resonance driving terms, the azimuthal harmonics of the sextupole distribution. Only the outgoing arms of the separatrix are shown in this figure. The figure must be oriented in such a way that the septum kick transforms into a displacement at the second septum while preserving the distinction between

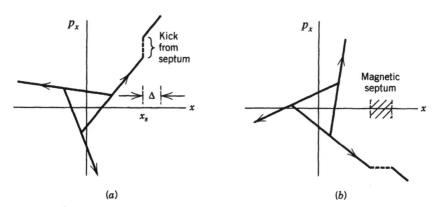

Figure 4.6. Phase space separatrices at (a) the location of the first (electrostatic) septum, and (b) the location of the second (magnetic) septum.

particles departing the aperture and those remaining in the accelerator. For instance, the phase space at the second septum might appear as shown in Figure 4.6(b). The 90° rotation between the two figures projects the kick of the first septum fully onto the second septum; the orientation of the first ensures that the extracted beam is well displaced from the circulating beam in the second.

All that remains to be specified are the strength of the resonance driving terms—the integrals A and B above—and the tune difference, $\delta\nu$, from resonance. Since the ratio of A and B is already set by the orientation of the separatrices at one of the septa, only two quantities remain to be determined. There are two conditions to be satisfied. The step size Δ should span the septum aperture for efficiency, and the stable area should correspond to the emittance of the beam at the onset of extraction. The relationships between the step size and stable area on the one hand and the driving terms and tune difference on the other were exhibited in the last section. A bit of geometry needs to be added to that discussion, having to do with the projection of motion along the separatrix onto the x-axis. The steps are not hard to carry out, if the critical reader wants to pursue them.

More in line with the focus of this discussion is the subject of extraction efficiency. As noted in the preceding paragraph, it was evident from the last section how to produce a slow spill beam. But what fraction of the particles strike the first septum?

In order to estimate the extraction inefficiency, let us suppose that the extraction process proceeds so slowly that it may be considered a static process. Then, the particle density distribution along an outgoing separatrix or along the projection of the separatrix onto a coordinate axis varies inversely as the rate of change of position along that coordinate. To convince oneself of this, let $F(x_1)\Delta x_1$ be the number of particles in an interval Δx_1 at x_1. After some time interval T has elapsed, the particles find themselves in

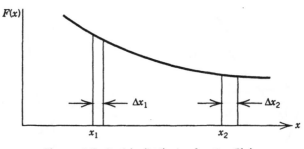

Figure 4.7. Particle distribution function $F(x)$.

Δx_2 at x_2. The number of particles is the same, so

$$F(x_2)\Delta x_2 = F(x_1)\Delta x_1. \qquad (4.54)$$

From

$$T = \int_{x_1}^{x_2} \frac{dx}{dx/dt} = \int_{x_1+\Delta x_1}^{x_2+\Delta x_2} \frac{dx}{dx/dt} \qquad (4.55)$$

it follows that

$$\int_{x_1}^{x_1+\Delta x_1} \frac{dx}{dx/dt} = \int_{x_2}^{x_2+\Delta x_2} \frac{dx}{dx/dt}, \qquad (4.56)$$

or

$$\frac{\Delta x_1}{(dx/dt)_1} = \frac{\Delta x_2}{(dx/dt)_2} \qquad (4.57)$$

where Figure 4.7 may be helpful in identifying the quantities.

So Δx varies directly as dx/dt; therefore $F(x)$ varies inversely as dx/dt. It will be more convenient to use the *turn number*, n, as the independent variable; that is, we take the spatial dependence of F to be of the form

$$F \propto \frac{1}{dx/dn}. \qquad (4.58)$$

If a septum of thickness w in the x-coordinate is located at a distance x_s from the central orbit, then the inefficiency e, defined as the fraction of particles that strike the first septum, is

$$e = \left[\int_{x_s}^{x_s+w} \frac{dx}{dx/dn}\right] \Big/ \left[\int_{x_s}^{\infty} \frac{dx}{dx/dn}\right] \qquad (4.59)$$

$$= \left[\int_{x_s}^{x_s+w} \frac{dx}{dx/dn}\right] \Big/ \left[\int_{x_s}^{x_s+\Delta} \frac{dx}{dx/dn}\right]. \qquad (4.60)$$

The second form above acknowledges, in the denominator, that the particle density distribution cuts off at a distance $x_s + \Delta$, with Δ being the *step size*, that is, the growth during x in the number of turns, N, between successive encounters with the septum at the proper phase for exit from the ring. In the third-integer case, $N = 3$.

The septum thickness w is small compared with x_s, and the integral in the numerator can be replaced by $w/(dx/dn)$ evaluated at x_s. The integral in the denominator is just N. So for the inefficiency we can use either of the following forms:

$$e = \frac{w}{(dx/dn)_{x_s}} \frac{1}{\int_{x_s}^{x_s+\Delta} dx/(dx/dn)} = \frac{1}{N} \frac{w}{(dx/dn)_{x_s}}. \qquad (4.61)$$

Note that the flatter the distribution F, the better the efficiency. This circumstance favors the choice of low order multipoles to generate the step size.

Let us estimate the inefficiency for third-integer extraction in the limit of vanishing stable phase space; the algebra is simplified by going to this limit, but all the principles remain the same. The situation is illustrated in Figure 4.8. Projected on the x-axis, the equation of motion for a particle traveling outward on the separatrix is

$$\frac{dx}{dn} = \frac{1}{4} \frac{\text{resonance driving term}}{\cos\theta} x^2. \qquad (4.62)$$

The expression for the inefficiency immediately gives

$$e = w \frac{x_s + \Delta}{x_s \Delta}. \qquad (4.63)$$

The expression in the numerator is related to the maximum displacement to

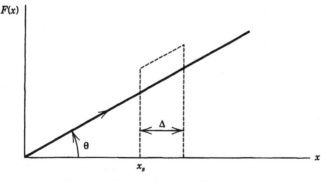

Figure 4.8. Step size across the septum.

be tolerated elsewhere in the ring, according to

$$x_s + \Delta = x_{max} \left(\frac{\beta_s}{\beta_0} \right)^{1/2} \cos \theta, \qquad (4.64)$$

where β_s is the amplitude function at the septum, and β_0 is the maximum value of the amplitude function in the standard cells of the ring where the aperture limitations are to be found. The angle θ has already been set by the arguments concerning the orientation of the separatrices at the two septa. For fixed $x_s + \Delta$, the mimimum in the inefficiency occurs for $x_s = \Delta$, and so

$$e_{min} = \frac{4w}{x_s + \Delta}. \qquad (4.65)$$

As a numerical example, let's take the superconducting synchrotron at Fermilab. The maximum oscillation amplitude in the arcs of the ring was fixed at 20 mm, after extensive simulation of particle motion in the fields provided by the superconducting main magnets. The lattice insertions for extraction devices were designed with β_s a factor of 2.3 larger than the maximum amplitude function in the standard cells. With $\theta = 45°$ and a septum thickness of 0.1 mm, the minimum inefficiency is 1.9%.

The accelerator in this example actually uses half-integer extraction; instead of sextupoles, the ring contains appropriate quadrupoles and octopoles. The distinction between the two approaches need not concern us here. The analogous calculation to that of the preceding paragraph leads to an inefficiency of 1.7%. It is interesting that inefficiency at this level implies generation of secondaries in the septum with a flux almost two orders of magnitude higher than would be tolerated by the superconducting magnets located downstream, requiring additional protective measures to be taken.

4.2.5 Comments on Correction Systems

In Chapter 3, we mentioned two types of correction magnets—those to compensate steering errors or make steering adjustments, and those to adjust the tune. These are elements of the correction magnet system. More properly this collection of elements should be called the correction and adjustment system, since it actually performs both functions.

Steering correction is conceptually simple. Tune correction brings with it the additional complexity of amplitude function perturbation, or equivalently, half-integer resonance excitation. Chromaticity compensation through the use of sextupole magnets can excite any of the sextupole driven resonances; or, if we wished to excite the third-integer resonance for slow extraction, we certainly would not wish to affect the chromaticity of the accelerator. That is, correction systems should perform specific functions cleanly without the introduction of possibly undesirable side effects.

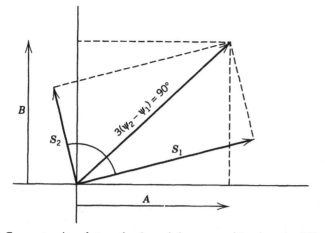

Figure 4.9. Two sextupoles of strengths S_1 and S_2 separated in phase by 90° on the $3\nu_0$ harmonic, where ν_0 is the third-integer resonance tune at which compensation is to be performed.

Since we have been discussing sextupole effects, let's concentrate on this case. For definiteness, suppose that we wish to compensate the one-degree-of-freedom third-integer resonance arising from field imperfections related to nonvanishing b_2 in the bending magnets of a synchrotron. This description is completely equivalent to the generation of resonance driving terms for third-integer extraction. In general, there will be two driving terms A and B, as given in Equations 4.39 and 4.40, generated by the errors. So we need two "knobs" which can be adjusted to compensate these driving terms. Note that the driving terms A and B are 90° out of phase with each other on the $3\nu_0$ harmonic.

The simplest way of effecting this correction might appear to be the introduction of two correction sextupoles at locations of equal values of the amplitude function β, differing in phase from one another by 90° on this harmonic; that is, the two correctors could be some odd multiple of 30° in betatron phase apart. We could represent these two sextupoles in a phasor diagram as shown in Figure 4.9, where the phasor amplitude is proportional to the strength of the sextupole, $S = B''l/2(B\rho)$.

The summation of all the sextupole field errors around the ring also can be represented by a phasor in this diagram. By a suitable choice of strengths and polarities of the two correction sextupoles, their resultant phasor can be made to cancel the phasor due to the errors, and hence the resonance driving terms given by Equations 4.39 and 4.40 can be brought to zero.

But unless we are so lucky that $\beta_1 S_1 D_1 = -\beta_2 S_2 D_2$, where D is the dispersion function, then the chromaticity would be changed by the correction. So, for pure resonance correction, we usually are led to a somewhat more complicated scheme. Suppose all of our harmonic compensation set are

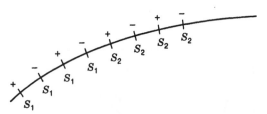

Figure 4.10. The $\nu_x = 19\frac{1}{3}$ resonance compensation circuit of the Tevatron contains sets of eight sextupoles, each located at a standard focusing quadrupole location in the FODO arc lattice, with the polarities shown. The sextupoles labeled S_1 are powered together, as are the ones labeled S_2.

located at equal values of dispersion as well. Then a scheme in which each sextupole is paired with a sextupole of equal strength but opposite sign 180° out of phase on the $3\nu_0$ harmonic will produce the desired results.

Let's carry this discussion to a more complicated case where the phase advance is not ideally suited to the introduction of sextupoles for resonance compensation. This is the circumstance, for instance, of the Tevatron. The phase advance per cell is 68°, and the tune is somewhat above $19\frac{1}{3}$. For the $19\frac{1}{3}$ resonance, the harmonic of interest is $k = 58$. Because the Tevatron is basically twofold symmetric, placement of equal strength sextupoles diametrically opposite one another will guarantee that only even harmonic resonances will be driven (or compensated).

The distribution of sextupoles to drive or compensate the $\nu_x = 19\frac{1}{3}$ resonance is shown in Figure 4.10. The amplitude function is the same at each sextupole, and the resulting phasor diagram is shown in Figure 4.11.

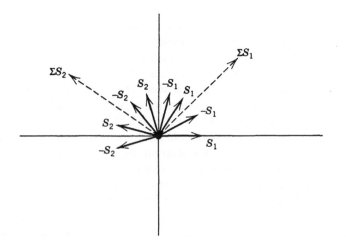

Figure 4.11. Phasor diagram resulting from set of eight sextupoles used to correct the $\nu_x = 19\frac{1}{3}$ resonance in the Tevatron.

The two families of sextupoles produce resultants which are approximately, though not exactly, 90° in phase apart on the $3\nu_0$ harmonic. Figure 4.10 shows two neighboring families of four sextupoles each, and in the Tevatron this pattern in replicated three additional times. The necessity for 32 sextupole magnets as opposed to, say, four is dictated by the need to produce a desired compensation strength while only limited space was available for an individual element.

It is probably clear that there is not a unique design approach for correction and adjustment systems. The approach reflects both the needs of the accelerator and the predelictions of the designer.

4.3 THE HAMILTONIAN FORMALISM

Only the most basic methods of dynamics have been used thus far, because we feel that the physics at work is most transparently illustrated in that way. But much of accelerator physics makes use of one form or another of higher dynamics. The Hamiltonian approach is the method most frequently encountered in the literature.[1]

In this section, we review the Hamiltonian form of dynamics, and then recast much of the material of the earlier discussions in this language.[2] No new physics is introduced, but the generality obtained may be helpful to the reader who wishes to pursue this approach further.

4.3.1 Review of Hamiltonian Dynamics

For a system with n degrees of freedom, there is a function $H(q, p, t)$ called the Hamiltonian. The variables are n generalized coordinates, their n conjugate momenta, and the time t. For the present, we are suppressing the subscripts on the variables, but we will include them when clarity demands it. The $2n$ equations of motion—Hamilton's equations—are then

$$\dot{p} = -\frac{\partial H}{\partial q}, \qquad \dot{q} = \frac{\partial H}{\partial p}. \tag{4.66}$$

The Hamiltonian approach focuses from the outset on motion in the $2n$-dimensional phase space of the dynamical variables p and q. At the beginning of our discussion, the variables will indeed resemble the momenta and coordinates of elementary mechanics. But that resemblance will fade as

[1]See, for example, R. D. Ruth, "Single-Particle Dynamics in Circular Accelerators," and L. Michelotti, "Introduction to the Nonlinear Dynamics Arising from Magneic Multipoles," *Physics of Particle Accelerators* (SLAC Summer School 1985, Fermilab Summer School 1984), AIP Conf. Proc. 153, 1987.
[2]H. Goldstein, *Classical Mechanics*, 2nd ed., Addison-Wesley Publishing Co., 1981.

we progress. In basic mechanics, we are all familiar with point transformations in configuration space. That is, we introduce new coordinates Q related to the old positions q by n equations of the form $Q = Q(q)$. In phase space, more general transformations among all $2n$ variables are possible and useful. All we require is that the form of Hamilton's equations be preserved.

Suppose we transform from variables p, q to variables P, Q, and that the new Hamiltonian is $K(P, Q, t)$. Hamilton's equations will be valid in both sets of coordinates, provided both satisfy the modified Hamilton's principle:

$$\delta \int (p_i \dot{q}_i - H) \, dt = 0, \tag{4.67}$$

$$\delta \int (P_i \dot{Q}_i - K) \, dt = 0, \tag{4.68}$$

where summation over the repeated indices is implied. The "modified" means that both positions and momenta are varied independently between the end points. The above will be satisfied if the integrands differ by only the total time derivative of some function F:

$$(p_i \dot{q}_i - H) = (P_i \dot{Q}_i - K) + \frac{dF}{dt}. \tag{4.69}$$

The transformations that maintain the validity of Hamilton's equations are called canonical transformations, and F is called the generating function. Note that the modified Hamilton's principle will remain valid also in the case that the integrands are in the ratio of some constant factor λ:

$$\lambda(p_i \dot{q}_i - H) = P_i \dot{Q}_i - K. \tag{4.70}$$

This is also a canonical transformation, corresponding to a scale change of the variables. We will encounter instances of this type of transformation in the next section.

The function F is in general a function of both the old and new variables as well as the time. We will restrict ourselves to functions that contain half of the old variables and half the new; these are useful for determining the explicit form of the transformation. The function F may then take on any of the following four forms:

$$F = F_1(q, Q, t), \tag{4.71}$$

$$F = F_2(q, P, t) - Q_i P_i, \tag{4.72}$$

$$F = F_3(Q, p, t) + q_i p_i, \tag{4.73}$$

$$F = F_4(p, P, t) + q_i p_i - Q_i P_i. \tag{4.74}$$

Now, if we insert each of these into

$$(p_i \dot{q}_i - H) = (P_i \dot{Q}_i - K) + \frac{dF}{dt}, \tag{4.75}$$

we obtain the relationships between old and new quantities listed below.

$$p = \frac{\partial F_1}{\partial q}, \qquad P = -\frac{\partial F_1}{\partial Q}, \tag{4.76}$$

$$p = \frac{\partial F_2}{\partial q}, \qquad Q = \frac{\partial F_2}{\partial P}, \tag{4.77}$$

$$q = -\frac{\partial F_3}{\partial p}, \qquad P = -\frac{\partial F_3}{\partial Q}, \tag{4.78}$$

$$q = -\frac{\partial F_4}{\partial p}, \qquad Q = \frac{\partial F_4}{\partial P}. \tag{4.79}$$

In all four cases,

$$K = H + \frac{\partial F_i}{\partial t}. \tag{4.80}$$

4.3.2 The Hamiltonian for Small Transverse Oscillations

The relativistic Hamiltonian for a particle of charge e moving under the influence of an electromagnetic field characterized by vector and scalar potentials $\vec{A}$ and V is

$$\mathscr{H} = \sqrt{\left(\vec{p} - e\vec{A}\right)^2 c^2 + m^2 c^4} + eV, \tag{4.81}$$

where $\vec{p}$ is the momentum conjugate to the Cartesian position coordinates of the particle. The magnetic field $\vec{B}$ and the electric field $\vec{E}$ are given by

$$\vec{B} = \nabla \times \vec{A}, \tag{4.82}$$

$$\vec{E} = -\nabla V - \frac{\partial \vec{A}}{\partial t}. \tag{4.83}$$

Recall that the canonical momentum $\vec{p}$ is related to the kinematic momentum $\gamma m \vec{v}$ by

$$\gamma m \vec{v} = \vec{p} - e\vec{A}. \tag{4.84}$$

In Chapter 3, we developed an equation of motion for small transverse deviations from the reference orbit. We want to follow the prescriptions of the Hamiltonian formalism and arrive at the same point. First, perform a canonical transformation with generating function

$$F = F_3(p, Q, t) = \vec{p} \cdot (\rho\hat{s} + x\hat{x} + y\hat{y}). \tag{4.85}$$

Then, following the rules of the preceding section,

$$p_s = \frac{\partial F_3}{\partial s} = \vec{p} \cdot \hat{s}\left(1 + \frac{x}{\rho}\right), \tag{4.86}$$

$$p_x = \frac{\partial F_3}{\partial x} = \vec{p} \cdot \hat{x}, \tag{4.87}$$

$$p_y = \frac{\partial F_3}{\partial y} = \vec{p} \cdot \hat{y}. \tag{4.88}$$

Again, as in Chapter 3, we assume that the curvature is locally a constant to simplify the discussion. Note that p_s is not the tangential component of the conjugate momentum. In order to preserve the relationship between components of the momentum and components of the vector potential, we define a canonical vector potential according to

$$A_s = \vec{A} \cdot \hat{s}\left(1 + \frac{x}{\rho}\right), \tag{4.89}$$

$$A_x = \vec{A} \cdot \hat{x}, \tag{4.90}$$

$$A_y = \vec{A} \cdot \hat{y}. \tag{4.91}$$

The generating function does not contain the time explicitly, so the new Hamiltonian, $\mathscr{H}'$, is just the old Hamiltonian expressed in the new coordinates:

$$\mathscr{H}' = c\left[\frac{1}{(1 + x/\rho)^2}(p_s - eA_s)^2 + (p_x - eA_x)^2 + (p_y - eA_y)^2 + m^2c^2\right]^{1/2}$$
$$+ eV \tag{4.92}$$

Now we want to change the independent variable from t to s. Consider $x' \equiv dx/ds$:

$$x' \equiv \frac{dx}{ds} = \frac{dx/dt}{ds/dt} = \frac{\partial \mathscr{H}/\partial p_x}{\partial \mathscr{H}/\partial p_s}. \tag{4.93}$$

The last form may be transformed into a partial derivative at constant $\mathcal{H}$ using

$$d\mathcal{H} = \left(\frac{\partial \mathcal{H}}{\partial p_x}\right)_{p_s} dp_x + \left(\frac{\partial \mathcal{H}}{\partial p_s}\right)_{p_x} dp_s = 0, \qquad (4.94)$$

or

$$\left(\frac{\partial p_s}{\partial p_x}\right)_{\mathcal{H}} = -\frac{(\partial \mathcal{H}/\partial p_x)_{p_s}}{(\partial \mathcal{H}/\partial p_s)_{p_x}}. \qquad (4.95)$$

Then for x' we have

$$x' = \left(\frac{\partial(-p_s)}{\partial p_x}\right)_{\mathcal{H}, p_y, x, y, s}. \qquad (4.96)$$

This has the form of a Hamilton's equation for x' with $-p_s$ playing the role of the Hamiltonian. If the same procedure is carried out for the entire set of Hamilton's equations, we find

$$x' = \frac{\partial H}{\partial p_x}, \qquad y' = \frac{\partial H}{\partial p_y}, \qquad t' = -\frac{\partial H}{\partial \mathcal{H}}, \qquad (4.97)$$

$$p_x' = -\frac{\partial H}{\partial x}, \qquad p_y' = -\frac{\partial H}{\partial y}, \qquad \mathcal{H}' = \frac{\partial H}{\partial t}. \qquad (4.98)$$

Therefore, the new pairs of canonical variables are x, p_x; y, p_y, and $t, -\mathcal{H}$, with the new Hamiltonian $H = -p_s$. Solving for p_s, we obtain for our new Hamiltonian

$$H = -p_s$$

$$= -\sqrt{\left[\left(\frac{\mathcal{H} - eV}{c}\right)^2 - m^2c^2 - (p_x - eA_x)^2 - (p_y - eA_y)^2\right]\left(1 + \frac{x}{\rho}\right)^2}$$

$$- eA_s. \qquad (4.99)$$

To proceed we consider, as before, only the case where the electric field is zero and where the magnetic field may be described by

$$\vec{B} = B_x(x, y)\hat{x} + B_y(x, y)\hat{y} \qquad (4.100)$$

for which we need only a single nonvanishing component of the vector potential, $\vec{A} \cdot \hat{s} \equiv A_z$. Therefore, we may write

$$H = -\sqrt{\left(\frac{\mathscr{H}^2}{c^2} - m^2c^2 - p_x^2 - p_y^2\right)\left(1 + \frac{x}{\rho}\right)^2} - eA_s. \quad (4.101)$$

For constant energy, $\mathscr{H} = E$ and thus

$$\frac{\mathscr{H}^2 - m^2c^4}{c^2} = p^2, \quad (4.102)$$

which gives us

$$\begin{aligned}
H &= -\sqrt{\left(p^2 - p_x^2 - p_y^2\right)\left(1 + \frac{x}{\rho}\right)^2} - eA_s \\
&= -p\left(1 + \frac{x}{\rho}\right)\sqrt{1 - \left(\frac{p_x}{p}\right)^2 - \left(\frac{p_y}{p}\right)^2} - eA_s \\
&\approx -p\left[1 + \frac{x}{\rho} - \frac{1}{2}\left(\frac{p_x}{p}\right)^2 - \frac{1}{2}\left(\frac{p_y}{p}\right)^2\right] - eA_s. \quad (4.103)
\end{aligned}$$

Finally, we must consider the form of A_s. From the definition of the vector potential and from Maxwell's equations, we find

$$\vec{B} = \vec{\nabla} \times \vec{A} = \frac{\partial A_z}{\partial y}\hat{x} - \frac{\partial A_z}{\partial x}\hat{y}, \quad (4.104)$$

$$\vec{\nabla} \times \vec{B} = \hat{s}\nabla^2 A_z = 0 \quad (4.105)$$

$$\Rightarrow \quad A_z = \mathrm{Re}\left\{\sum_n \frac{B^{(n)}}{n!}(x + iy)^n\right\}, \quad (4.106)$$

or

$$A_z = -B_0 - \tfrac{1}{2}B'(x^2 - y^2) - \tfrac{1}{6}B''(x^3 - 3xy^2) - \cdots. \quad (4.107)$$

Thus, A_s may be obtained from

$$A_s = \vec{A} \cdot \hat{s}\left(1 + \frac{x}{\rho}\right). \quad (4.108)$$

Scaling the Hamiltonian and the conjugate momentum variable by a constant, namely the design momentum p_0,

$$H \rightarrow H/p_0, \tag{4.109}$$

$$p_x \rightarrow p_x/p_0, \tag{4.110}$$

the final form of the Hamiltonian becomes

$$H = -\left(1 + \frac{x}{\rho} - \tfrac{1}{2}p_x^2 - \tfrac{1}{2}p_y^2\right)$$

$$+ \left(\frac{eB_0}{p_0}x + \frac{1}{2}\frac{eB'}{p_0}(x^2 - y^2)\right)\left(1 + \frac{x}{\rho}\right) + \cdots . \tag{4.111}$$

We may now apply Hamilton's equations to generate the equations of motion:

$$x' = \frac{\partial H}{\partial p_x} = p_x, \tag{4.112}$$

$$p_x' = -\frac{\partial H}{\partial x} = \left(\frac{1}{\rho} - \frac{eB_0}{p_0}\right) - \frac{eB_0}{p_0\rho}x - \frac{eB'}{p_0}x$$

$$= \left(\frac{1}{\rho} - \frac{1}{p_0}\right) - \left(\frac{1}{\rho p_0} + \frac{eB'}{p_0}\right)x, \tag{4.113}$$

which, for the ideal momentum particle ($\rho = p_0$), becomes

$$p_x' = x'' = -\left(\frac{1}{\rho_0^2} + \frac{eB'}{p_0}\right)x, \tag{4.114}$$

which is the same as the result of Chapter 3.

4.3.3 Transformations of the Hamiltonian

In one degree of freedom, our small oscillation Hamiltonian is of the form

$$H = \frac{x'^2}{2} + \frac{K(s)x^2}{2} + \cdots . \tag{4.115}$$

The first two terms resemble the Hamiltonian for a simple harmonic oscillator, though with the typical variation of the spring constant with s. In the last section, we found that a phase-amplitude description of the motion was useful. In the Hamiltonian formalism, the corresponding quantities are called action-angle variables. We wish to identify them and carry out the appropriate canonical transformation. The resulting Hamiltonian will still be a function of s, and a further transformation will be employed to remove that dependence. At that point, the "unperturbed" Hamiltonian will be a constant of the motion. One could, of course, invert the order of these transformations and arrive at the same point.

The form of the first transformation is suggested by the solutions that we have already developed. Recall, for the linear motion, that

$$x = \mathscr{A}\sqrt{\beta}\,\cos{(\psi + \delta)}, \tag{4.116}$$

$$\beta x' + \alpha x = -\mathscr{A}\sqrt{\beta}\,\sin(\psi + \delta). \tag{4.117}$$

The Hamiltonian H and the solution above suggest that the new unperturbed Hamiltonian, H_0, should be of the form

$$H_0 = H_0(\mathscr{A}) = \text{constant.} \tag{4.118}$$

So we are motivated to select ψ as the coordinate and look for a conjugate variable, J, which is related to the amplitude. That is,

$$x = \mathscr{A}(J)\sqrt{\beta}\,\cos{\psi}, \tag{4.119}$$

$$\beta x' + \alpha x = -\mathscr{A}(J)\sqrt{\beta}\,\sin{\psi}, \tag{4.120}$$

where the arbitrary constant δ has been absorbed into the definition of ψ.

We can easily express x' in terms of x and ψ:

$$x' = -\frac{x\tan{\psi} + \alpha x}{\beta}. \tag{4.121}$$

Therefore, we look for a generating function of the type $F_1(x, \psi, s)$, for which

$$x' = \frac{\partial F_1}{\partial x}, \qquad J = -\frac{\partial F_1}{\partial \psi}. \tag{4.122}$$

From the expression for x' above, integration with respect to x gives

$$F_1 = -\frac{\alpha + \tan{\psi}}{2\beta}x^2 + f(\psi), \tag{4.123}$$

where $f(\psi)$ is an arbitrary function which we may set equal to zero. Then,

$$J = -\frac{\partial F_1}{\partial \psi} = \frac{\sec^2{\psi}}{2\beta}x^2, \tag{4.124}$$

and from the expression for x in terms of $\mathscr{A}$ we see that

$$J = \frac{\mathscr{A}^2}{2}, \tag{4.125}$$

or

$$\mathscr{A}(J) = \sqrt{2J}. \tag{4.126}$$

The new "unperturbed" Hamiltonian becomes just

$$H_0 = \frac{J}{\beta(s)}. \tag{4.127}$$

Now we wish to remove the s-dependence. Let us choose a new dependent variable which advances linearly with s in the unperturbed problem (which ψ does not). The quantity

$$\int \frac{ds}{\beta} - 2\pi\nu\frac{s}{C} \tag{4.128}$$

represents the "flutter" of the phase with respect to the average phase advance. Here, C is the circumference of the accelerator. So we wish to define a new coordinate, θ, such that

$$\psi = \theta + \text{"flutter"} = \theta + \int \frac{ds}{\beta} - \nu\frac{s}{R}. \tag{4.129}$$

This expression contains the old and new coordinate variables. Therefore, an F_1 transformation is not appropriate. Let's try an F_2. We want

$$\theta = \psi + \nu\frac{s}{R} - \int \frac{ds}{\beta}, \tag{4.130}$$

$$I = J, \tag{4.131}$$

and

$$J = \frac{\partial F_2}{\partial \psi}, \qquad \theta = \frac{\partial F_2}{\partial I}. \tag{4.132}$$

Following the same reasoning as above, the generating function is

$$F_2 = I\left(\psi + \nu\frac{s}{R} - \int \frac{ds}{\beta}\right), \tag{4.133}$$

and the new unperturbed Hamiltonian is then

$$H_0 = \frac{\nu}{R} I. \tag{4.134}$$

4.3.4 The Third-Integer Resonance Revisited

Having found an appropriate form for the unperturbed Hamiltonian H_0, we may now treat the remaining terms of H as perturbations and thus investigate the effects of nonlinearities on the particle motion. Let us consider the next term in the expansion of the Hamiltonian, namely, that due to a sextupole field. We have

$$
\begin{aligned}
H &= \frac{1}{2}p_x^2 + \left(\frac{1}{2}\frac{eB'}{p_0}\right)x^2 + \frac{1}{6}\frac{eB''}{p_0}x^3 \\
&= \tfrac{1}{2}x'^2 + \tfrac{1}{2}K(s)x^2 + \tfrac{1}{3}S(s)x^3 \\
&= H_0 + \tfrac{1}{3}S(s)x^3.
\end{aligned} \tag{4.135}
$$

The transformations thus far have yielded the following relations:

$$x = \sqrt{2\beta I} \cos \chi, \tag{4.136}$$

$$x' = -\sqrt{\frac{2I}{\beta}} (\sin \chi + \alpha \cos \chi), \tag{4.137}$$

where

$$\chi \equiv \theta - \nu\frac{s}{R} + \int \frac{ds}{\beta}. \tag{4.138}$$

So the new Hamiltonian is

$$
\begin{aligned}
H &= \frac{\nu}{R}I + \tfrac{1}{3}S(s)(2\beta I)^{3/2} \cos^3\!\left(\theta - \nu\frac{s}{R} + \int \frac{ds}{\beta}\right) \\
&= H_0 + \tfrac{1}{3}S(s)(2\beta I)^{3/2} \cos^3\!\left(\theta - \nu\frac{s}{R} + \int \frac{ds}{\beta}\right).
\end{aligned} \tag{4.139}
$$

Now we are in a position to proceed in much the way that we did in the last section. The amplitude function and sextupole strength are periodic functions of s, so we may expand the factor containing them in a Fourier series:

$$\beta^{3/2}S(s) = \sum_m W_m \cos[m\phi(s)], \tag{4.140}$$

$$W_m \equiv \frac{1}{\pi R} \int \beta^{3/2}S(s) \cos(m\phi) \, ds, \tag{4.141}$$

where the angular variable ϕ is s/R, and for simplicity we write only the cosine terms. Then, also expanding the cosine-cubed term, the Hamiltonian becomes

$$
H = \frac{\nu}{R}I + \tfrac{1}{12}(2I)^{3/2} \sum_m W_m \cos m\phi (\cos 3\chi + 3\cos \chi)
$$

$$
= \frac{\nu}{R}I + \tfrac{1}{24}(2I)^{3/2} \sum_m W_m [\cos(3\chi + m\phi) + \cos(m\phi - 3\chi)
$$

$$
+ 3\cos(\chi + m\phi) + 3\cos(m\phi - \chi)]. \quad (4.142)
$$

From the equation for the rate of change of the action,

$$
\frac{dI}{ds} = -\frac{\partial H}{\partial \theta}
$$

$$
= \tfrac{1}{24}(2I)^{3/2} \sum_m W_m [-3\sin(3\chi + m\phi) + 3\sin(m\phi - 3\chi)
$$

$$
- 3\sin(\chi + m\phi) + 3\sin(m\phi - \chi)], \quad (4.143)
$$

we see that if there is an integer m such that $3\chi \approx m\phi$, then the condition for a resonance is satisfied. To examine the phase space near resonance, we go to rotating coordinates as we did in the more elementary treatment. Now, of course, this requires yet another canonical transformation. We take

$$
F_2 = I_1\left(\theta - \frac{\nu_0 s}{R}\right), \quad (4.144)
$$

where $\nu_0 = m/3$ is the resonant tune. This transformation leaves the initial action I and the final action I_1 the same, so we will suppress the subscript. The new coordinate, θ_1 is

$$
\theta_1 = \theta - \nu_0 \frac{s}{R} \quad (4.145)
$$

with the new Hamiltonian given by

$$
H = \frac{\delta}{R}I + \tfrac{1}{24}(2I)^{3/2}W_m \cos\left(m\phi - 3\theta_1 + 3\delta\,\phi - 3\int \frac{ds}{\beta}\right). \quad (4.146)
$$

As before, $\delta \equiv \nu - \nu_0$.

In these coordinates, Hamilton's equations give

$$
\frac{dI}{ds} = \tfrac{1}{8}(2I)^{3/2}W_m \sin\left(m\phi - 3\theta_1 + 3\delta\,\phi - 3\int \frac{ds}{\beta}\right), \quad (4.147)
$$

$$
\frac{d\theta_1}{ds} = \delta + \tfrac{1}{16}(2I)^{1/2}W_m \cos\left(m\phi - 3\theta_1 + 3\delta\,\phi - 3\int \frac{ds}{\beta}\right). \quad (4.148)
$$

At the fixed points, the derivatives above are each equal to zero. The first tells us that the argument of the cosine function in the Hamiltonian must be equal to an integer times π. From the second, the distance to the fixed points is given by

$$I = \left(\frac{2\delta}{W_m}\right)^2.$$
(4.149)

To the degree that I can be related to the amplitude $\mathscr{A}$ of the unperturbed motion according to the relationship used above, the amplitude of the fixed point is given by

$$\mathscr{A} = \frac{8\delta}{W_m}.$$
(4.150)

This is the same as the result in the last section, taking into account the different definitions of amplitudes and Fourier coefficients.

PROBLEMS

1. Derive the inhomogeneous equation of motion after the Floquet transformation has been applied (Equation 4.5).

2. For a picture frame dipole magnet where the cores meet perfectly on one side but are separated by a small gap h on the other, show that the quadrupole term generated is given by

$$b_1 = \frac{h}{gw},$$

where g is the nominal gap height and w is the pole width.

3. Consider a unit square in the tune diagram (i.e. ν_V vs. ν_H) with corners at (n, n), $(n + 1, n)$, $(n, n + 1)$, $(n + 1, n + 1)$. Draw the lines representing all sum resonances through fourth order.

4. Using Equations 4.48 and 4.49, find the first integral to the equation of motion as given in the text.

5. Integrate the equation of motion (Equation 4.52) along the vertical separatrix for the resonance considered in the text. Verify that the

number of turns, n, to progress from y_0 to y is

$$n = \frac{1}{2\sqrt{3}\,\pi\delta}\,\ln\left\{\frac{\dfrac{A}{4\pi\delta}y_0 + \sqrt{3}}{\dfrac{A}{4\pi\delta}y_0 - \sqrt{3}}\cdot\frac{\dfrac{A}{4\pi\delta}y - \sqrt{3}}{\dfrac{A}{4\pi\delta}y + \sqrt{3}}\right\}.$$

6. Assume that a single thin sextupole is placed in a ring. The point of observation is chosen to be at the midpoint of the sextupole. Its strength is such that the harmonic driving term A is

$$A \approx \beta\frac{B''L}{2(B\rho)} = 0.05 \text{ mm}^{-1}.$$

For positive δ, the separatrices will be oriented as sketched in the text above. Take $\delta = 0.006$. Using the result of the preceding problem, find the position after 3 turns of a particle that starts from $y_0 = 10$ mm.

7. The single sextupole case would appear to be a long way from the spirit of the derivation upon which the analytical results are based. Using a computer, carry out a turn by turn calculation for the same particle as that in the example above. That is, start at the midpoint of the sextupole and give the particle a deflection appropriate to half of the sextupole. Propagate around the ring with a linear matrix, then deliver another half-sextupole kick. Compare with the result of the preceding problem. How do you know if you are even on the separatrix?

8. The length scale and sextupole strength in Problem 6 are useful in relating the dynamics to realistic values in accelerators. But for calculational purposes, it is easier to cast the mappings in dimensionless form. Note that the driving term A of Problem 6 has the dimensions of inverse length. Using $1/A$ as the unit of length, introduce new variables with x and p_x scaled accordingly. State the mapping of the preceding problem in these new variables. Note that the only remaining parameter is the tune, and so phase space plots developed in these variables are characteristic of the tune only. Modify the program that you wrote for the preceding problem accordingly, and repeat the calculations. Compare the behavior near resonance ($\delta = 0.006$) with that far from resonance ($\delta = 0.09$).

9. Investigate the deformation of phase space from a circle as a single sextupole is turned on by an extension of the turn by turn calculation carried out in Problem 8. As the phase space occupied by the beam exceeds the stable region provided by the ring, observe that particles depart along or near the separatrices in the near resonance situation of

the static case provided the rate of variation of the sextupole strength is sufficiently slow. This is a multiparticle problem, and you will have to decide how the initial phase space is to be populated.

10. The field of an octopole magnet varies as the cube of the horizontal or vertical displacement from its center, so one might expect an average octopole moment to produce a fourth-power term in the "potential" for betatron oscillations. The oscillation frequency would then depend on the oscillation amplitude. Carry out the same steps as used in the derivation of the amplitude and phase equations but with octopole fields rather than sextupole fields. Assume that the tune is far from resonance, so that any harmonic driving terms are unimportant. Show that the equations of motion are

$$\frac{da}{dn} = 0,$$

$$\frac{d\psi}{dn} = \tfrac{3}{8}a^2 D$$

with

$$D \equiv \frac{\beta_0}{(B\rho)} \oint \left(\frac{\beta}{\beta_0}\right)^2 \frac{B'''}{6} \, ds.$$

The tune change with amplitude is then

$$\Delta\nu(a) = \frac{1}{2\pi}\frac{3}{8}Da^2.$$

11. Although the stopband width arising from quadrupole errors is a problem connected with the linear motion, we didn't calculate the width in Chapter 3. It is relatively easy to do at this point, as another example of the method used for treating nonlinear motion. Repeat the steps of the sextupole case, but with quadrupole terms instead. For simplicity, assume that the only nonvanishing integral is

$$Q = \frac{\beta_0}{(B\rho)} \oint \frac{\beta}{\beta_0} B' \cos 2\nu_0\phi \, ds.$$

Show that the equations of motion in x and p_x are

$$\frac{dx}{dn} = -\tfrac{1}{2}Qp_x + 2\pi\delta \cdot p_x,$$

$$\frac{dp_x}{dn} = -\tfrac{1}{2}Qx - 2\pi\delta \cdot x,$$

where δ is now the difference in tune from the half integer. Show that the motion is unstable for all particles over the tune range of the unperturbed ring given by

$$\Delta\nu = \frac{1}{2\pi}|Q|.$$

This is the stopband width.

12. The results of Problem 10 and Problem 11 may be combined to illustrate another approach to slow extraction—the so-called half-integer method. Suppose the unperturbed tune is just below one-half of an integer. In the presence of an average octopole moment, some large amplitude particles may find themselves in the half-integer stopband if the appropriate quadrupole harmonic is present. Show that, in lowest order of perturbation theory, the separatrices in this case consist of two intersecting circles.

CHAPTER 5

Transverse Coupled Motion

In this chapter we generalize the discussion of transverse oscillations to two degrees of freedom. From a design point of view it would be convenient if the two transverse motions were independent. But in a real accelerator, such is not the case. A horizontal betatron oscillation will be deflected into the vertical degree of freedom if the particle encounters a horizontal magnetic field transverse to its direction of motion. Suppose in our otherwise perfectly constructed accelerator a quadrupole magnet has been rotated through a small angle ϕ, clockwise looking along the beam direction, so that the poles are no longer inclined at angles of $45°$, but rather at angles $45° \pm \phi$. Then the transverse components of the magnetic field that the particle encounters will be

$$B_y = B'(x \cos 2\phi + y \sin 2\phi),$$
$$B_x = B'(y \cos 2\phi - x \sin 2\phi),$$

(5.1)

where B' is the gradient of the quadrupole. (See Problem 1 at the end of the chapter.) Then a particle which starts out with an oscillation in the horizontal plane will see a vertical deflection when it passes through this field. Similarly, a vertical oscillation will receive a horizontal deflection. This is an example of linear coupling, in that the equations of the coupled motion remain linear in x and y. An example of nonlinear coupling is that due to a normal sextupole field:

$$B_x = B''xy,$$
$$B_y = \tfrac{1}{2}B''(x^2 - y^2).$$

(5.2)

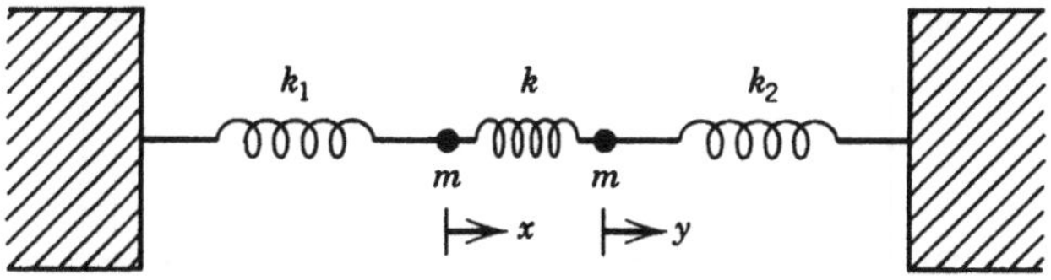

Figure 5.1. A system of two masses and three springs.

Here, a pure horizontal oscillation will remain so, while an oscillation which begins solely in the vertical will couple into the horizontal degree of freedom.

In the following two sections we examine linear and nonlinear coupling, respectively. Though we do not attempt an exhaustive treatment of either, we hope to expose the essential physics of the topic.

5.1 LINEAR COUPLING

As an introduction to coupled motion we will review the familiar case of a pair of coupled simple harmonic oscillators. Then we extend the perturbation formalism developed in the previous chapter to the problem of coupling induced by rotated quadrupole fields. We conclude with remarks on the 4×4 matrix formulation and treatment.

5.1.1 Coupled Harmonic Oscillators

Suppose we have a spring system such as is shown in Figure 5.1. The equations of motion are

$$m\ddot{x} + (k_1 + k)x - ky = 0, \tag{5.3}$$

$$m\ddot{y} + (k_2 + k)y - kx = 0, \tag{5.4}$$

where k_1, k_2, and k are the restoring forces per unit displacement for springs 1 and 2 and for the coupling spring, respectively. If we let $\omega_1^2 \equiv (k_1 + k)/m$, $\omega_2^2 \equiv (k_2 + k)/m$, and $q^2 \equiv k/m$, then the equations of motion become

$$\ddot{x} + \omega_1^2 x - q^2 y = 0, \tag{5.5}$$

$$\ddot{y} + \omega_2^2 y - q^2 x = 0. \tag{5.6}$$

Following the standard procedure of identifying normal mode frequencies, we look for particular solutions of the form

$$x = e^{i\omega t}, \tag{5.7}$$

$$y = ae^{i\omega t}. \tag{5.8}$$

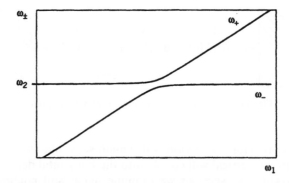

Figure 5.2. Eigenfrequencies of system of two coupled harmonic oscillators.

Substituting these solutions into the equations of motion, we find

$$\omega_1^2 - \omega^2 = aq^2, \tag{5.9}$$

$$a(\omega_2^2 - \omega^2) = q^2, \tag{5.10}$$

and thus

$$(\omega_1^2 - \omega^2)(\omega_2^2 - \omega^2) = q^4. \tag{5.11}$$

Solving for ω^2 we find

$$\omega^2 = \frac{\omega_1^2 + \omega_2^2 \pm \sqrt{(\omega_2^2 - \omega_1^2)^2 + 4q^4}}{2}. \tag{5.12}$$

Though there seem to be four frequencies, they occur in positive and negative pairs; we use the positive solutions below. A plot of the two frequencies as functions of ω_1 (keeping ω_2 constant) is shown in Figure 5.2. The figure illustrates that for ω_1 much less than ω_2, the eigenfrequency associated with mass 1 is ω_-; for ω_1 much greater than ω_2, the associated eigenfrequency is ω_+. When the unperturbed frequencies are near each other, one can no longer associate one frequency with each mass. Rather, the motion is composed of a superposition of the two eigenfrequencies.

Figure 5.2 also illustrates a minimum separation of the two eigenfrequencies when ω_1 is near ω_2. For the special case where $\omega_1 = \omega_2 = \omega_0$ (i.e., where $k_1 = k_2 = k_0$),

$$\omega = \sqrt{\omega_0^2 \pm q^2}$$

$$= \sqrt{k_0/m}, \quad \sqrt{(k_0 + 2k)/m}. \tag{5.13}$$

Notice that one of the frequencies is that of the unperturbed motion. This is the frequency for the mode in which the two masses oscillate in phase with each other and thus the coupling spring exerts no force.

The two frequencies of the system cannot be the same. Even if the frequencies of the two uncoupled oscillators were identical, the degeneracy would be broken by the introduction of coupling. The smallest frequency separation ($\Delta\omega$) can be found by writing the two eigenfrequencies as

$$\omega_+^2 = \omega_0^2 + q^2, \tag{5.14}$$

$$\omega_-^2 = \omega_0^2 - q^2. \tag{5.15}$$

Thus

$$\omega_+^2 - \omega_-^2 = (\omega_+ + \omega_-)(\omega_+ - \omega_-) = 2q^2, \tag{5.16}$$

and for $\omega_+ \approx \omega_- \approx \omega_0$, the minimum angular frequency difference is

$$\Delta\omega = q^2/\omega_0. \tag{5.17}$$

The more detailed treatments of the remaining subsections examine other facets of the coupled oscillator problem with particular application to particle motion in magnetic fields. But the practical essentials are contained in the foregoing simple discussion. Coupling moves the frequencies about—moves the betatron oscillation tunes, in the case of an accelerator—and so can defeat the precise tune control needed to avoid resonances in devices such as colliders. At an even more elementary level, coupling is an irritant in diagnosing beam behavior, for the eigenfrequencies and eigenmodes are no longer associated with the degrees of freedom specified in the design.

5.1.2 Perturbation Treatment of a Single Skew Quadrupole

In high energy accelerators, there are two sources of linear coupling insofar as single particle motion is concerned: rotated (or *skew*) quadrupole fields arising from alignment errors, and occasionally, solenoidal fields associated with high energy physics experiments. Here we concentrate on the first, though the techniques readily admit of generalization.

We apply the method of variation of parameters used in Chapter 4 to the motion in the presence of a single skew quadrupole. In the introduction to this chapter, we wrote down the field of a quadrupole which had rotated about its axis through an angle ϕ from its normal orientation. For small ϕ, the changes in x' and y' for a particle traversing this rotated quadrupole, in the thin lens approximation, are

$$\Delta x' = -\frac{B_y l}{(B\rho)} = -q(x + 2\phi y),$$ (5.18)

$$\Delta y' = \frac{B_x l}{(B\rho)} = q(y - 2\phi x),$$ (5.19)

where x and y are the transverse betatron oscillation displacements at entry to the quadrupole, and q is the inverse focal length of the thin lens quadrupole. In matrix form, these equations become

$$\begin{pmatrix} x \\ x' \\ y \\ y' \end{pmatrix} = \begin{pmatrix} 1 & 0 & 0 & 0 \\ -q & 1 & -2\phi q & 0 \\ 0 & 0 & 1 & 0 \\ -2\phi q & 0 & q & 1 \end{pmatrix} \begin{pmatrix} x \\ x' \\ y \\ y' \end{pmatrix}_0 .$$ (5.20)

The matrix above can be factored into the product of the matrix of a normal quadrupole and the matrix of a pure skew quadrupole (that is, one which has its poles rotated by 45°) of strength $2\phi q$:

$$M = M_{sq} M_q = M_q M_{sq},$$ (5.21)

where

$$M_q = \begin{pmatrix} 1 & 0 & 0 & 0 \\ -q & 1 & 0 & 0 \\ 0 & 0 & 1 & 0 \\ 0 & 0 & q & 1 \end{pmatrix},$$ (5.22)

$$M_{sq} = \begin{pmatrix} 1 & 0 & 0 & 0 \\ 0 & 1 & -2\phi q & 0 \\ 0 & 0 & 1 & 0 \\ -2\phi q & 0 & 0 & 1 \end{pmatrix}.$$ (5.23)

Thus we see that the rotated quadrupole can be treated as a pure skew element perturbing the basic lattice.

As in Chapter 4, we look at the long term effect of a single perturbing element, in this case a skew quadrupole, in a synchrotron. We will find that if the sum of the unperturbed tunes is an integer, we have an unstable resonance; if the difference is an integer, we have stable but coupled motion, and we will obtain an expression for the minimum tune separation in the latter case, analogous to the corresponding expression obtained for the classical coupled oscillator derived above.

Let's describe the horizontal and vertical motion at the location of the pertubation by

$$x = a\sqrt{\frac{\beta_x}{\beta_0}} \cos \psi_x, \tag{5.24}$$

$$p_x = \alpha_x x + \beta_x x' \tag{5.25}$$

$$= -a\sqrt{\frac{\beta_x}{\beta_0}} \sin \psi_x, \tag{5.26}$$

$$y = b\sqrt{\frac{\beta_y}{\beta_0}} \cos \psi_y, \tag{5.27}$$

$$p_y = \alpha_y y + \beta_y y' = -b\sqrt{\frac{\beta_y}{\beta_0}} \sin \psi_y. \tag{5.28}$$

In this form, a and b are the amplitudes of the motion as measured at a common value of the amplitude function β_0. Again, in the thin element approximation we obtain $\Delta x = \Delta y = 0$ in passing through the skew quadrupole and

$$\Delta \psi_x = \frac{\Delta a}{a} \cot \psi_x, \tag{5.29}$$

$$\Delta \psi_y = \frac{\Delta b}{b} \cot \psi_y. \tag{5.30}$$

Substituting $\Delta x'$ and $\Delta y'$ due to the skew quadrupole, we obtain

$$\Delta a = k\sqrt{\beta_x \beta_y} \, b \sin \psi_x \cos \psi_y, \tag{5.31}$$

$$\Delta b = k\sqrt{\beta_x \beta_y} \, a \sin \psi_y \cos \psi_x, \tag{5.32}$$

$$\Delta \psi_x = k\sqrt{\beta_x \beta_y} \, \frac{b}{a} \cos \psi_x \cos \psi_y, \tag{5.33}$$

$$\Delta \psi_y = k\sqrt{\beta_x \beta_y} \, \frac{a}{b} \cos \psi_x \cos \psi_y, \tag{5.34}$$

where $k = 2\phi q$ for our particular case of a rotated quadrupole.

We may use trigonometric identities to rewrite the difference equations. Looking at the first,

$$\Delta a = \tfrac{1}{2} k\sqrt{\beta_x \beta_y} \, b\big[\sin(\psi_x + \psi_y) + \sin(\psi_x - \psi_y)\big], \tag{5.35}$$

we see that there are two possibilities for secular change, namely, if the difference between the horizontal and vertical tunes is near an integer or if their sum is near an integer. Consider the difference equations for the amplitudes. The situation of the preceding sentence leads to the following relationship for fractional change in the two amplitudes:

$$\frac{\Delta a}{b} = \pm \frac{\Delta b}{a}, \tag{5.36}$$

where the upper sign goes with the case where the sum is near an integer, and the lower sign for the other case. An invariant of the motion follows from integration of the above:

$$a^2 + b^2 = \text{constant}, \qquad \nu_x - \nu_y = \text{integer}; \tag{5.37}$$

$$a^2 - b^2 = \text{constant}, \qquad \nu_x + \nu_y = \text{integer}. \tag{5.38}$$

For the first case, the *difference resonance*, the sum of the squares of the amplitudes is constant and therefore the motion is bounded. In contrast, for the *sum resonance* the amplitudes lie on hyperbolae in *a-b* space, and hence the amplitudes may become arbitrarily large.

We will now concentrate on the particular case of the difference resonance. We must again transform our difference equations into differential equations. As earlier, the amplitude equations can be readily converted. However, the change in the phase advance per turn is not small, and so a more suitable angular variable must be introduced.

Suppose the tunes ν_x and ν_y are a distance δ away from the line $\nu_x - \nu_y = m$ in tune space as indicated in Figure 5.3. Then we may define new variables $\chi_x = \psi_x - 2\pi(\nu_0 + m)$, $\chi_y = \psi_y - 2\pi\nu_0$, and the new differ-

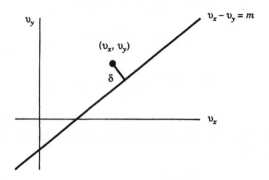

Figure 5.3. Operating point near a difference resonance.

ential equations we wish to solve are

$$\frac{da}{dn} = \tfrac{1}{2}\kappa b \sin(\chi_x - \chi_y), \tag{5.39}$$

$$\frac{db}{dn} = -\tfrac{1}{2}\kappa a \sin(\chi_x - \chi_y), \tag{5.40}$$

$$\frac{d\chi_x}{dn} = \tfrac{1}{2}\kappa \frac{b}{a} \cos(\chi_x - \chi_y) + \pi\sqrt{2}\,\delta, \tag{5.41}$$

$$\frac{d\chi_y}{dn} = \tfrac{1}{2}\kappa \frac{a}{b} \cos(\chi_x - \chi_y) - \pi\sqrt{2}\,\delta, \tag{5.42}$$

where $\kappa \equiv k\sqrt{\beta_x\beta_y} = 2\phi q\sqrt{\beta_x\beta_y}$. These equations are more easily solved in Cartesian coordinates. Because we have transformed to a new angular coordinate, we define new rectangular coordinates, which we still refer to as x and y to keep the nomenclature simple. Since

$$\frac{dx}{dn} = \frac{x}{a}\frac{da}{dn} + p_x\frac{d\chi_x}{dn}, \tag{5.43}$$

$$\frac{dp_x}{dn} = \frac{p_x}{a}\frac{da}{dn} - x\frac{d\chi_x}{dn}, \tag{5.44}$$

and similarly for dy/dn and dp_y/dn, we may generate the equations of motion in the new coordinates. Expanding the trigonometric functions

$$\cos(\chi_x - \chi_y) = \cos\chi_x \cos\chi_y + \sin\chi_x \sin\chi_y \tag{5.45}$$

$$= \left(\frac{x}{a}\right)\left(\frac{y}{b}\right) + \left(\frac{-p_x}{a}\right)\left(\frac{-p_y}{b}\right), \tag{5.46}$$

$$\sin(\chi_x - \chi_y) = \sin\chi_x \cos\chi_y - \sin\chi_y \cos\chi_x \tag{5.47}$$

$$= \left(\frac{-p_x}{a}\right)\left(\frac{y}{b}\right) - \left(\frac{-p_y}{b}\right)\left(\frac{x}{a}\right), \tag{5.48}$$

the equations of motion become

$$\frac{dx}{dn} = \frac{\kappa}{2}\frac{1}{x^2 + p_x^2}\left[x(xp_y - yp_x) + p_x(xy + p_xp_y)\right] + \sqrt{2}\,\pi\delta p_x, \tag{5.49}$$

$$\frac{dp_x}{dn} = \frac{\kappa}{2}\frac{1}{x^2 + p_x^2}\left[p_x(xp_y - yp_x) - x(xy + p_xp_y)\right] - \sqrt{2}\,\pi\delta x, \tag{5.50}$$

$$\frac{dy}{dn} = \frac{\kappa}{2}\frac{1}{y^2 + p_y^2}\left[y(xp_y - yp_x) - p_y(xy + p_xp_y)\right] - \sqrt{2}\,\pi\delta p_y, \tag{5.51}$$

$$\frac{dp_y}{dn} = \frac{\kappa}{2}\frac{1}{y^2 + p_y^2}\left[p_y(xp_y - yp_x) + y(xy + p_xp_y)\right] + \sqrt{2}\,\pi\delta y. \tag{5.52}$$

The various bracketed expressions hint that it would be useful to introduce complex variables of the form

$$u = x + ip_x, \tag{5.53}$$

$$v = y + ip_y, \tag{5.54}$$

in terms of which the above equations can be reduced to

$$\frac{du}{dn} = -i\frac{\kappa}{2}v - i\pi\sqrt{2}\,\delta u, \tag{5.55}$$

$$\frac{dv}{dn} = -i\frac{\kappa}{2}u + i\pi\sqrt{2}\,\delta v. \tag{5.56}$$

We are now at the point where we can solve two coupled equations, as we did in Section 5.1.1, for the eigenfrequencies. By using the trial solutions

$$u = e^{i2\pi v n}, \tag{5.57}$$

$$v = w e^{i2\pi v n} \tag{5.58}$$

we find

$$2\pi v = -\frac{\kappa}{2}w - \pi\sqrt{2}\,\delta, \tag{5.59}$$

$$w(2\pi v) = -\frac{\kappa}{2} + \pi\sqrt{2}\,\delta w. \tag{5.60}$$

Solving for w yields

$$w = \frac{-4\pi\delta\sqrt{2} \pm \sqrt{\left(4\pi\delta\sqrt{2}\right)^2 + 4\kappa^2}}{2\kappa}, \tag{5.61}$$

and substitution into Equation 5.59 gives the two eigentunes in the rotating coordinate system:

$$v_\pm = \pm\frac{1}{8\pi}\sqrt{\left(4\pi\delta\sqrt{2}\right)^2 + 4\kappa^2}. \tag{5.62}$$

So for the tune difference we have

$$\Delta v \equiv v_+ - v_- = \sqrt{2\delta^2 + (\kappa/2\pi)^2}. \tag{5.63}$$

If we are exactly on resonance, with $v_x + m = v_y = v_0$, we see that the minimum tune difference is

$$\Delta \nu = \frac{|\kappa|}{2\pi} = \frac{|k|}{2\pi}\sqrt{\beta_x \beta_y} = \frac{|\phi q|}{\pi}\sqrt{\beta_x \beta_y}. \tag{5.64}$$

This is the same separation of the fractional parts of the tunes that would be observed after we transformed out of the rotating coordinate system. In fact, a standard procedure for measuring the coupling strength in a synchrotron is to attempt to make the fractional parts of the horizontal and vertical tunes identical, generating a curve similar to that in Figure 5.2.

In Chapter 4, we carried out the third-integer resonance calculation for the general case of a distribution of arbitrarily many elements. Here, we leave the many element case to the exercises at the end of the chapter. The main feature added by the many element case is analogous to that of the earlier treatment. Resonances are driven by particular azimuthal harmonics of the perturbations; in the case of the linear difference resonance, there are two such driving terms. Thus, two corrector sets are the minimum needed to compensate a nearby linear coupling resonance.

5.1.3 Matrix Treatment of a Single Skew Quadrupole

It is instructive to approach the same problem of the previous subsection using matrix algebra. In the perturbative treatment we can obtain complete analytical expressions describing the motion of the particle, but the solution is of necessity approximate. In particular, the eigenvalues are not exact. By recasting the problem in matrix language, we can find the exact eigenvalues at the expense of giving up the elementary closed form solutions for the motion.

As suggested in Equation 5.24, the 4×4 matrix for a thin skew quadrupole can be written in the form

$$M_{sq} = \begin{pmatrix} 1 & 0 & 0 & 0 \\ 0 & 1 & -k & 0 \\ 0 & 0 & 1 & 0 \\ -k & 0 & 0 & 1 \end{pmatrix}, \tag{5.65}$$

where k is the skew quadrupole strength. Suppose our unperturbed matrix for one turn through the synchrotron is

$$\begin{pmatrix} A & 0 \\ 0 & B \end{pmatrix} = \begin{pmatrix} a & b & 0 & 0 \\ c & d & 0 & 0 \\ 0 & 0 & e & f \\ 0 & 0 & g & h \end{pmatrix}, \tag{5.66}$$

where A and B are 2×2 matrices. Then the matrix for the perturbed accelerator at the point just after the skew quadrupole will be

$$
M_{sq}\begin{pmatrix} A & 0 \\ 0 & B \end{pmatrix} = \begin{pmatrix} a & b & 0 & 0 \\ c & d & -ek & -fk \\ 0 & 0 & e & f \\ -ak & -bk & g & h \end{pmatrix}. \tag{5.67}
$$

We find the eigenvalues by the usual method, i.e., $\det(M - \lambda I) = 0$. We find

$$
\left(\lambda + \frac{1}{\lambda} - \text{Tr}\,A\right)\left(\lambda + \frac{1}{\lambda} - \text{Tr}\,B\right) - bfk^2 = 0. \tag{5.68}
$$

Solving for $\lambda + 1/\lambda$,

$$
\lambda + \frac{1}{\lambda} = \frac{\text{Tr}\,A + \text{Tr}\,B}{2} \pm \frac{\sqrt{(\text{Tr}\,A - \text{Tr}\,B)^2 + 4bfk^2}}{2}. \tag{5.69}
$$

As in Chapter 3, we set $\lambda = e^{i\mu}$, where μ may be complex. Then,

$$
2\cos\mu = \cos\mu_x + \cos\mu_y \pm \sqrt{(\cos\mu_x - \cos\mu_y)^2 + k^2\beta_x\beta_y \sin\mu_x \sin\mu_y}. \tag{5.70}
$$

Here, we have written the elements of the matrices A and B in terms of the unperturbed Courant-Snyder parameters β_x and β_y at the location of the skew quadrupole, and the unperturbed phase advances μ_x and μ_y. We have, as promised, arrived at the exact expression for the eigenfrequencies. The result reduces to that of the previous subsection in the appropriate limit. For example, let us obtain the expression for the minimum tune separation, i.e., when $\cos\mu_x = \cos\mu_y$. Setting

$$
\mu_+ = \mu_0 - \frac{2\pi\,\Delta\nu}{2}, \tag{5.71}
$$

$$
\mu_- = \mu_0 + \frac{2\pi\,\Delta\nu}{2}, \tag{5.72}
$$

where $\Delta\nu$ is small, then

$$
2\cos\mu_+ - 2\cos\mu_- \approx 2\pi\,\Delta\nu \sin\mu_0 = |k|\sqrt{\beta_x\beta_y}\,\sin\mu_0, \tag{5.73}
$$

or

$$
\Delta\nu = \frac{|k|}{2\pi}\sqrt{\beta_x\beta_y}, \tag{5.74}
$$

which is the identical result displayed in Equation 5.64.

A numerical example is in order. Remember that $k = 2\phi q$, where ϕ is the rotation angle of a quadrupole with inverse focal length q. For a typical FODO lattice, $q\sqrt{\beta_x\beta_y} \approx 2$ at a quadrupole location. Suppose $\phi = 10^{-2}$ (10 mrad). Then $\Delta\nu \approx 0.006$. If there were a large number N of such perturbations, one would expect $\Delta\nu$ to scale like $\sqrt{N}$. For a Tevatron scale synchrotron, $N \approx 200$ and $\sqrt{N}\,\Delta\nu \approx 0.1$—much too large to assure low order resonance avoidance. So the typical tolerance on ϕ is at the 10^{-3} level.

5.1.4 Matrix Formalism of Linear Coupling

In our treatment of transverse motion in one degree of freedom, the 2×2 matrices describing the motion had three independent parameters because the matrix propagating motion from one point to another had unit determinant. The 4×4 matrices for two-degree of freedom coupled motion also have unit determinants, which itself assures that not all 16 matrix elements are independent. But there is an even more restrictive condition that is in fact the proper generalization of the one-degree of freedom result. We will find that there are six relationships among the 16 matrix elements.

Let's go back to one degree of freedom and write Hamilton's equations in matrix form, which we can do for a linear system. Equation 4.115, $\mathcal{H} = p_x^2/2 + K_x x^2/2$, can be written in matrix form as

$$\mathcal{H} = \frac{1}{2}(x \quad p_x)\begin{pmatrix} K_x & 0 \\ 0 & 1 \end{pmatrix}\begin{pmatrix} x \\ p_x \end{pmatrix} \equiv \tfrac{1}{2}\vec{X}^T H \vec{X}, \tag{5.75}$$

where $\vec{X}^T$ is the transpose of $\vec{X}$. Since $x' = \partial\mathcal{H}/\partial p_x$ and $p_x' = -\partial\mathcal{H}/\partial x$, then we may write the equations of motion as

$$\vec{X}' = \begin{pmatrix} 0 & 1 \\ -K_x & 0 \end{pmatrix}\vec{X} = \begin{pmatrix} 0 & 1 \\ -1 & 0 \end{pmatrix}\begin{pmatrix} K_x & 0 \\ 0 & 1 \end{pmatrix}\vec{X} \equiv SH\vec{X}. \tag{5.76}$$

Now, suppose $\vec{X}_1$ and $\vec{X}_2$ are two independent solutions to this differential equation. The bilinear form $p_{x1}x_2 - x_1 p_{x2} = \vec{X}_2^T S \vec{X}_1$, called the Wronskian, is a well-known invariant. To see this, we evaluate its derivative:

$$\left(\vec{X}_2^T S \vec{X}_1\right)' = \vec{X}_2'^T S \vec{X}_1 \vec{X}_2^T S \vec{X}_1' \tag{5.77}$$

$$= \left(SH\vec{X}_2\right)^T S \vec{X}_1 + \vec{X}_2^T S\left(SH\vec{X}_1\right) \tag{5.78}$$

$$= \vec{X}_2^T H^T S^T S \vec{X}_1 + \vec{X}_2^T SSH\vec{X}_1 \tag{5.79}$$

$$= \vec{X}_2^T H \vec{X}_1 - \vec{X}_2^T H \vec{X}_1 \tag{5.80}$$

$$= 0, \tag{5.81}$$

where we have made use of the conditions $H^T = H$, $S^T S = I$, and $S^2 = -I$.

This result is not limited to one degree of freedom. The Hamiltonian matrix is in general symmetric. The bilinear form,

$$p_{x1}x_2 - x_1 p_{x2} + p_{y1}y_2 - y_1 p_{y2} + \cdots \tag{5.82}$$

extends the Wronskian form to n degrees of freedom and is an invariant. The generalization of the matrix S to n degrees of freedom is

$$\begin{pmatrix} 0 & 1 & & & \\ -1 & 0 & & & \\ & & 0 & 1 & \\ & & -1 & 0 & \\ & & & & \ddots \end{pmatrix}, \tag{5.83}$$

and the proof is formally identical to the one degree of freedom case.

From this invariant we may derive a condition on the matrices describing the evolution of the motion. For suppose M propagates the vector $\vec{X}$ between two points in the accelerator s_1 and s_2. Then,

$$\vec{X}_1(s_2) = M\vec{X}_1(s_1), \tag{5.84}$$

$$\vec{X}_2(s_2) = M\vec{X}_2(s_1), \tag{5.85}$$

and so

$$\vec{X}_2^T(s_2)S\vec{X}_1(s_2) = \vec{X}_2^T(s_1)M^TSM\vec{X}_1(s_1). \tag{5.86}$$

But

$$\vec{X}_2^T(s_2)S\vec{X}_1(s_2) = \vec{X}_2^T(s_1)S\vec{X}_1(s_1), \tag{5.87}$$

which gives us the so-called *symplectic condition* which M must satisfy:

$$M^TSM = S. \tag{5.88}$$

Here, the symplectic condition has been derived for a linear Hamiltonian system. This is a special case of a general result that can be stated as follows: The Jacobian matrix of a canonical transformation is symplectic.[1] So even if we were considering nonlinear motion in a Hamiltonian system, the matrix describing motion in the neighborhood of a particular trajectory would be symplectic.

The symplectic condition gives $n(2n - 1)$ independent relationships among the $2n \times 2n$ elements of an n-degree of freedom matrix. For our case, $n = 2$; thus there are six relationships among the matrix elements. To see

[1]See, for example, L. A. Pars, *A Treatise on Analytical Dynamics*, Wiley, New York, 1965, p. 497.

this, we write the 4×4 matrix in the form

$$M = \begin{pmatrix} A & E \\ F & B \end{pmatrix}, \tag{5.89}$$

where the elements of M as written above are themselves 2×2 matrices. Setting $M^T S M$ equal to S, we find the four matrix relationships

$$A^T S A + F^T S F = S, \tag{5.90}$$

$$A^T S E + F^T S B = 0, \tag{5.91}$$

$$E^T S A + B^T S F = 0, \tag{5.92}$$

$$E^T S E + B^T S B = S. \tag{5.93}$$

If one writes out the first and last equations one finds they are equivalent to

$$\det A + \det F = 1, \tag{5.94}$$

$$\det E + \det B = 1, \tag{5.95}$$

and so the apparent eight relationships implied by these two matrix equations in fact reduce to only two. The second and third matrix equations are in fact the same: one is just the transpose of the other. So from these, we have four additional relationships for a total of six. One could extend the Courant-Snyder approach to express the matrix M in terms of $16 - 6 = 10$ independent parameters.[2] Rather, we content ourselves with some general conclusions.

For the remainder of this discussion M will denote the single-turn matrix. For each eigenvalue of this symplectic matrix, its reciprocal is also an eigenvalue. We can demonstrate this by letting λ_1 and $\vec{X}_1$ be an eigenpair and λ_k, $\vec{X}_k$ be one of the remaining eigenpairs. Then,

$$\left(M\vec{X}_k\right)^T S M \vec{X}_1 = \vec{X}_k^T \lambda_k S \lambda_1 \vec{X}_1 = \vec{X}_k^T S \vec{X}_1, \tag{5.96}$$

where the equality of the first and third expressions reflects the invariant of Equation 5.82. So the second and third expressions tell us that

$$(\lambda_1 \lambda_k - 1)\left(\vec{X}_k^T S \vec{X}_1\right) = 0. \tag{5.97}$$

But $\vec{X}_k^T S \vec{X}_1$ cannot vanish for all possible remaining choices of $\vec{X}_k$, because the eigenvectors form a basis. Since $\vec{X}_1^T S \vec{X}_1$ is zero, $S\vec{X}_1$ is orthogonal to $\vec{X}_1$. But $S\vec{X}_1$ can be expanded in terms of the eigenvectors, and so that expansion

---

[2]D. A. Edwards, and L. C. Teng, "Parametrization of Linear Coupled Motion in Periodic Systems," IEEE Trans. Nucl. Sci. NS-20, No. 3 (1973).

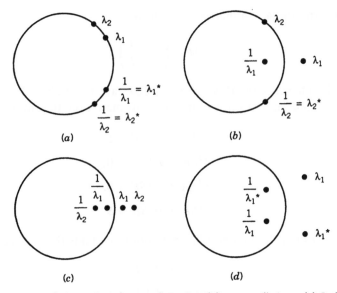

Figure 5.4. Location of eigenvalues for two-dimensional linear oscillations. (a) Both modes stable. (b) One mode stable, one mode unstable. (c) Both modes unstable in absence of coupling. (d) Instability induced by coupling. From Courant and Snyder (see Bibliography) with permission.

must contain at least one of the others. Therefore, there is at least one eigenvector for which $\vec{X}_k^T S \vec{X}_1$ is not zero, and for that k, $\lambda_k = 1/\lambda_1$. Hence the conclusion that eigenvalues occur in reciprocal pairs.

We already know that if one eigenvalue is complex then its conjugate is also an eigenvalue, because the coefficients of the eigenvalue equation are real. Therefore, the eigenvalues will occur in one of the four configurations shown in Figure 5.4.

Our goal now is to obtain the general solution to the eigenvalue equation. Note that if λ is an eigenvalue, then not only is $1/\lambda$ also an eigenvalue of M but also of M^{-1}. Therefore, it suffices to solve for the quantity $\Lambda \equiv \lambda + 1/\lambda$, which is an eigenvalue of $M + M^{-1}$. It helps to define the symplectic conjugate

$$\bar{A} \equiv -SA^T S. \tag{5.98}$$

If A is a 2×2 matrix, then

$$\bar{A} = \begin{pmatrix} A_{22} & -A_{12} \\ -A_{21} & A_{11} \end{pmatrix}. \tag{5.99}$$

If A is nonsingular, then we recognize that $\bar{A} = A^{-1} \det A$. With this definition, the symplectic conjugate of our 4×4 matrix M is

$$\bar{M} = -SM^TS = \begin{pmatrix} \bar{A} & \bar{F} \\ \bar{E} & \bar{B} \end{pmatrix}. \tag{5.100}$$

Since for our case $\det M = 1$, we have $\bar{M} = M^{-1}$ and so we want to find the eigenvalues Λ of

$$M + M^{-1} = M + \bar{M} = \begin{pmatrix} A + \bar{A} & E + \bar{F} \\ F + \bar{E} & B + \bar{B} \end{pmatrix}. \tag{5.101}$$

Using $A + \bar{A} = (\operatorname{Tr} A)I$, the characteristic equation is

$$\begin{vmatrix} (\operatorname{Tr} A - \Lambda)I & E + \bar{F} \\ F + \bar{E} & (\operatorname{Tr} B - \Lambda)I \end{vmatrix} = 0, \tag{5.102}$$

which reduces to

$$I\Lambda^2 - (\operatorname{Tr} A + \operatorname{Tr} B)I\Lambda + (\operatorname{Tr} A)(\operatorname{Tr} B)I - (F + \bar{E})(E + \bar{F}) = 0. \tag{5.103}$$

But

$$(F + \bar{E})(E + \bar{F}) = I \det(\bar{E} + F) \tag{5.104}$$

and so

$$\Lambda^2 - (\operatorname{Tr} A + \operatorname{Tr} B)\Lambda + (\operatorname{Tr} A)(\operatorname{Tr} B) - \det(\bar{E} + F) = 0. \tag{5.105}$$

This gives us the eigenfrequencies

$$2\cos 2\pi\nu = \frac{\operatorname{Tr} A + \operatorname{Tr} B}{2} \pm \frac{\sqrt{(\operatorname{Tr} A - \operatorname{Tr} B)^2 + 4\det(\bar{E} + F)}}{2}. \tag{5.106}$$

In general, $\operatorname{Tr} A$ and $\operatorname{Tr} B$ should not be confused with $2\cos \mu_x$ and $2\cos \mu_y$; their values will actually vary when computed for various points in the lattice. The interplay among the elements of the matrices A, B, E, and F are such that, of course, the eigenvalues we have found are independent of the point at which the matrices are computed.

5.2 NONLINEAR COUPLING

The main point of this section is to extend the treatment of Chapter 4 to identify the resonance lines in the ν_x, ν_y tune plane. The discussion of the $3\nu_x$ = integer resonance showed that there are trajectories extending to infinity in the phase plane; here, we show this for two-degree-of-freedom resonances in general.

But first, as an aid in finding the path toward the general result, we look at a particular coupling resonance in some detail in the next subsection.

5.2.1 Two-Degree-of-Freedom Sum Resonance Due to Distribution of Sextupoles

In Chapter 4, it was shown how a sextupole field can excite a resonance of the form $3\nu_x = P$, where P is an integer. It was mentioned that tune values which obey relationships of the form $2\nu_x + \nu_y = P$, $\nu_x + 2\nu_y = P$, and $3\nu_y = P$ are also to be avoided. Here we will examine one such case, $\nu_x + 2\nu_y = P$, in detail for a distribution of sextupole fields around the accelerator.

The normal sextupole field

$$B_y = \frac{B''}{2}(x^2 - y^2), \tag{5.107}$$

$$B_x = B''xy \tag{5.108}$$

can excite the sum resonance $\nu_x + 2\nu_y = P$ as will be confirmed below. We wish to examine the effect of the resonance on a neighboring line in the ν_x, ν_y diagram characterized by

$$\nu_x + 2\nu_y = P + 3\delta. \tag{5.109}$$

As we have done before, we let the x and y displacements at some particular point of observation in the lattice be written as

$$x = a\left(\frac{\beta_x}{\beta_0}\right)^{1/2} \cos \psi_x, \tag{5.110}$$

$$y = b\left(\frac{\beta_y}{\beta_0}\right)^{1/2} \cos \psi_y. \tag{5.111}$$

where β_x and β_y are the amplitude functions at the point of observation, ψ_x and ψ_y the phases, and β_0 is some convenient normalization factor. If, at the observation point, $\beta_x = \beta_y$, then it would be natural to take β_0 equal to β_x and β_y. Since it is not generally the case that the amplitude functions are the same at the position chosen, the value used for β_0 is arbitrary—actually, it is only introduced so that the amplitudes a and b will have the dimensions of length.

On the turn around the ring following passage of the observation point with amplitudes a and b, and phases ψ_x and ψ_y, we write

$$x = a\left(\frac{\beta_x(s)}{\beta_0}\right)^{1/2}\cos\Psi_x, \quad\quad \Psi_x \equiv \psi_x + \nu_x\phi_x, \quad\quad (5.112)$$

$$y = b\left(\frac{\beta_y(s)}{\beta_0}\right)^{1/2}\cos\Psi_y, \quad\quad \Psi_y \equiv \psi_y + \nu_x\phi_y, \quad\quad (5.113)$$

where ϕ_x and ϕ_y are the "reduced" phases ($\phi \equiv \psi/\nu$) which run from 0 to 2π in one turn. For a turn, the contributions to the change of amplitude and phase due to the sextupoles are accumulated in the approximation that the amplitudes remain constant and the phases develop as in the linear motion.

For passage through an element of length Δs in a sextupole, we have

$$\Delta x = \Delta y = 0, \quad\quad\quad\quad\quad\quad\quad\quad (5.114)$$

$$\Delta x' = -\frac{B''}{2(B\rho)}(x^2 - y^2)\,\Delta s, \quad\quad\quad (5.115)$$

$$\Delta y' = \frac{B''}{(B\rho)}xy\,\Delta s. \quad\quad\quad\quad\quad (5.116)$$

The changes in amplitude and phase are related to those in displacement and slope by

$$\Delta a = -(\beta_x\beta_0)^{1/2}\sin\Psi_x\,\Delta x', \quad\quad\quad (5.117)$$

$$\Delta\psi_x = \frac{\Delta a}{a}\frac{\cos\Psi_x}{\sin\Psi_x}, \quad\quad\quad\quad\quad (5.118)$$

with corresponding expressions in the y-plane. Inserting $\Delta x'$ from above, we have

$$\Delta a = (\beta_x\beta_0)^{1/2}\frac{B''}{2(B\rho)}\sin\Psi_x\left[a^2\left(\frac{\beta_x}{\beta_0}\right)\cos^2\Psi_x - b^2\left(\frac{\beta_y}{\beta_0}\right)\cos^2\Psi_y\right]\Delta s.$$

$$(5.119)$$

The first term in the brackets will be ignored, since it is involved only in the one-dimensional x-plane resonances. So we have

$$\Delta a = -\frac{B''\beta_0}{2(B\rho)}\left(\frac{\beta_x^{1/2}\beta_y}{\beta_0^{3/2}}\right)b^2\sin\Psi_x\cos^2\Psi_y\,\Delta s. \qquad (5.120)$$

If the trigonometric functions are expanded and only terms relating to the $2\nu_x + \nu_y$ resonance are retained, we have

$$\sin\Psi_x\cos^2\Psi_y \to \tfrac{1}{2}\sin\Psi_x\cos 2\Psi_y \to \tfrac{1}{4}\sin(\Psi_x + 2\Psi_y)$$

$$= \tfrac{1}{4}\big[\sin(\psi_x + 2\psi_y)\cos(\nu_x\phi_x + 2\nu_y\phi_y)$$

$$+ \cos(\psi_x + 2\psi_y)\sin(\nu_x\phi_x + 2\nu_y\phi_y)\big]. \qquad (5.121)$$

Integrating around the accelerator, we have for the change in x-amplitude per turn

$$\frac{da}{dn} = -\frac{b^2}{4}\big[A\sin(\psi_x + 2\psi_y) + B\cos(\psi_x + 2\psi_y)\big], \qquad (5.122)$$

where

$$A \equiv \oint \frac{B''\beta_0}{2(B\rho)}\left(\frac{\beta_x^{1/2}\beta_y}{\beta_0^{3/2}}\right)\cos\big[\nu_x\phi_x + 2\nu_y\phi_y\big]\,ds, \qquad (5.123)$$

$$B \equiv \oint \frac{B''\beta_0}{2(B\rho)}\left(\frac{\beta_x^{1/2}\beta_y}{\beta_0^{3/2}}\right)\sin\big[\nu_x\phi_x + 2\nu_y\phi_y\big]\,ds. \qquad (5.124)$$

Proceeding in the same way for $\Delta\psi_x$, the change in ψ_x due to the sextupoles is, for one turn,

$$\Delta\psi_x = -\frac{1}{4}\frac{b^2}{a}\big[A\cos(\psi_x + 2\psi_y) - B\sin(\psi_x + 2\psi_y)\big]. \qquad (5.125)$$

This expression for $\Delta\psi$ may be *provisionally* turned into a differential equation for the phase advance per turn by the addition of a term reflecting the phase change in the absence of sextupoles.

$$\frac{d\psi_x}{dn} = -\frac{1}{4}\frac{b^2}{a}\big[A\cos(\psi_x + 2\psi_y) - B\sin(\psi_x + 2\psi_y)\big] + 2\pi\delta_x. \qquad (5.126)$$

Not only is the added term $2\pi\delta_x$ not small (being $\approx 120°$ per turn), but also δ_x is not uniquely defined insofar as distance from the resonance in question is concerned. Hence the term "provisional"; when results are finally obtained, the arbitrariness concerning δ_x (and similarly δ_y) must be resolved.

The differential equations for the y-motion are obtained by the same procedure. Before summarizing the results, note that ψ_x and ψ_y only appear in the combination $\psi_x + 2\psi_y$ in da/dn and $d\psi_x/dn$; such is the case for the y-motion.

We define

$$\theta \equiv \psi_x + 2\psi_y. \tag{5.127}$$

Also, assume that only the A-terms are present. This may be the case as the result of symmetry, or the point of observation can be displaced until it is so. Then the differential equations are

$$\frac{da}{dn} = -\frac{b^2}{4}A \sin \theta, \tag{5.128}$$

$$\frac{db}{dn} = -\tfrac{1}{4}(2ab)A \sin \theta, \tag{5.129}$$

$$\frac{d\theta}{dn} = -\frac{1}{4}\left(4a + \frac{b^2}{a}\right)A \cos \theta + 6\pi\delta. \tag{5.130}$$

The equation for $d\theta/dn$ results from the addition of the equations for $d\psi_x/dn$ and $2\,d\psi_y/dn$. Whereas in the latter, δ_x and δ_y were not uniquely defined, in the equation for $d\theta/dn$, δ has a well-defined meaning, and for an operating point ν_x, ν_y close to the resonance, $6\pi\delta$ is small.

Division of the first equation by the second and integrating yields a constant of the motion

$$c = 2a^2 - b^2. \tag{5.131}$$

For fixed c, the amplitudes a and b can become arbitrarily large. Thus there are trajectories which extend to infinity. Eliminating b in favor of a and c from the first and third equations and integrating yields a second constant of the motion:

$$k = -A \cos \theta \, a^3 + \tfrac{1}{2}cA \cos \theta \, a + 2\pi\delta \cdot 3a^2. \tag{5.132}$$

The remainder of the discussion consists in the interpretation of these two constants of the motion.

For the following discussion, we express a and b in terms of a unit length $2\pi\delta/A$. The invariants, in these units, are

$$c = 2a^2 - b^2, \tag{5.133}$$

$$k = -\cos \theta \, a^3 + 3a^2 + \tfrac{1}{2}c \cos \theta \, a. \tag{5.134}$$

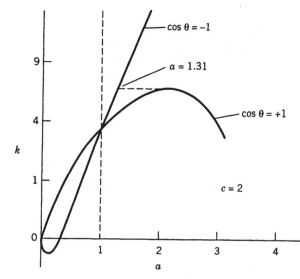

Figure 5.5. Constant of motion k vs. amplitude a for the particular case where $c = 2$. Here a has been normalized to $2\pi\delta / A$.

To see how the limits of stability may be inferred from these relations, consider the particular case $c = 2$. Then,

$$k = \cos\theta\,(a - a^3) + 3a^2. \tag{5.135}$$

The possible values of a (for this c) can be found by plotting k vs. a for $\cos\theta = \pm 1$. The result is shown in Figure 5.5. For $c = 2$, a must be > 1; otherwise b would be imaginary. The motion is stable if $3 \le k < 6.1$. This will be the case regardless of initial phase, θ, for all a such that $1 \le a \le \approx 1.31$. If $1.31 \le a \le 2.15$, the motion will be stable for some initial phases, unstable for others. If $a \ge 2.15$, the motion will be unstable regardless of the initial phase.

The value of a (call it a_1) for which $dk/da = 0$ on the $\cos\theta = 1$ curve can be found from

$$\frac{dk}{da} = -3a^2 + 6a + \frac{c}{2} = 0, \tag{5.136}$$

or,

$$a_1 = 1 \pm \left(1 + \frac{c}{6}\right)^{1/2}. \tag{5.137}$$

For positive c, the solution is single valued. Since we must also have $2a^2 \geq c$, it follows that for $c \geq 18$ ($a \geq 3$) no motion will be stable. For $c < 18$, any $a \geq a_1$ will be unstable. By replacing c with $2a^2 - b^2$ in the expression for a_1, a stability boundary in the a, b diagram is obtained:

$$a - 1 = \pm \left(1 + \frac{2a^2 - b^2}{6}\right)^{1/2},$$

(5.138)

or

$$4a^2 - 12a + b^2 = 0.$$

(5.139)

This expression defines a boundary outside of which all a and b are unstable, provided $a \geq 1$. For $a \leq 1$ a different boundary condition takes over, as discussed later. For the moment, we stay with $a \geq 1$.

For $a \leq a_1$, and given c, some motions will be stable and others unstable, depending on the initial θ. However, all a less than a value a_2 will be stable, where a_2 is defined by the condition that it lies on the $\cos\theta = -1$ line in Figure 5.5 with the same value of k as a_1. In the $c = 2$ example, $a_2 = 1.31$. That is,

$$a_2^3 + 3a_2^2 - \tfrac{1}{2}ca_2 = -a_1^3 + 3a_1^2 + \tfrac{1}{2}ca_1.$$

(5.140)

Expressing a_1 in terms of c, solving for a_2, and eliminating c in favor of b yields the line in the a, b plane beneath which all motion is stable:

$$\sqrt{2}\, b = 3 - a.$$

(5.141)

For $a \leq 1$, arbitrarily small a amplitudes can be unstable. Here, the stability boundary, beyond which all amplitudes in the a, b plane will be unstable is given by

$$\sqrt{2}\, b = 3 + a.$$

(5.142)

The regions of instability, partial stability, and complete stability are plotted in Figure 5.6.

It is interesting that one can go this far in analyzing a two-degree-of-freedom nonlinear resonance, and clearly we could pursue the discussion of this example considerably further. For our purposes a less detailed treatment is more suitable, and so we turn to the general discussion of the next subsection.

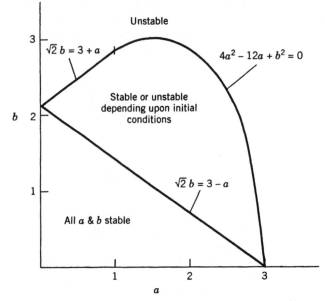

Figure 5.6. Stability regions in a, b plane for $\nu_x + 2\nu_y = N$ resonance. Here a and b are expressed in units of $2\pi\delta / A$.

5.2.2 Multipoles and Resonance Lines

In the opening section of Chapter 4 we stated without proof that all tune relationships of the form $M\nu_x + N\nu_y = P$ could lead to resonant behavior due to the multipole content of real magnetic fields. We are now in a position to prove this statement.

Let us start by developing a form for the general multipole expansion for the transverse field components in a bending magnet. We know that the solutions to Laplace's equation $\nabla^2 \phi_m = 0$, where ϕ_m is the magnetic scalar potential, have, in polar coordinates, the form

$$s_k r^k \sin k\theta, \tag{5.143}$$

$$c_k r^k \cos k\theta, \tag{5.144}$$

where s_k and c_k are constants. We can obtain the coefficients by working out a few particular cases. Here, we will take just the single example of a sextupole field, i.e., $k = 3$. Then, for the sine term,

$$B_y = \frac{\partial \phi_m}{\partial y} = 3s_3(x^2 - y^2), \tag{5.145}$$

$$B_x = \frac{\partial \phi_m}{\partial x} = 6s_3 xy. \tag{5.146}$$

Conforming to the notation used in Chapter 4, Equation 4.6, we identify s_3 with the normal sextupole coefficient b_2 according to $s_3 = B_0 b_2 / 3$.

The cosine term yields another contribution to the field:

$$B_y = \frac{\partial \phi_m}{\partial y} = -6c_3 xy, \qquad (5.147)$$

$$B_x = \frac{\partial \phi_m}{\partial x} = 3c_3 (x^2 - y^2). \qquad (5.148)$$

Note that B_x contains a term proportional to x^2; that is, in the $y = 0$ plane, the field points toward a pole located in that plane. This is the hallmark of a skew multipole. By analogy with the normal multipole coefficients b_n, we define skew coefficients, in this case the skew sextupole coefficient a_2, according to $c_3 = B_0 a_2 / 3$.

So, in general, our magnetic scalar potential may be written in the form

$$\phi_m = B_0 \sum_n \left[\frac{b_n}{n+1} r^{n+1} \sin(n+1)\theta + \frac{a_n}{n+1} r^{n+1} \cos(n+1)\theta \right], \qquad (5.149)$$

which can be recast in rectangular coordinates as

$$\phi_m = B_0 \sum_n \left[\frac{b_n}{n+1} \operatorname{Im}(x+iy)^{n+1} + \frac{a_n}{n+1} \operatorname{Re}(x+iy)^{n+1} \right], \qquad (5.150)$$

where $\operatorname{Im}(z)$ and $\operatorname{Re}(z)$ are the imaginary and real parts of the complex variable z. It is now straightforward to compute B_y and B_x:

$$B_y = \frac{\partial \phi_m}{\partial y} = B_0 \sum_n \left[b_n \operatorname{Re}(x+iy)^n - a_n \operatorname{Im}(x+iy)^n \right], \qquad (5.151)$$

$$B_x = \frac{\partial \phi_m}{\partial x} = B_0 \sum_n \left[b_n \operatorname{Im}(x+iy)^n + a_n \operatorname{Re}(x+iy)^n \right]. \qquad (5.152)$$

These results may be written in compact form:

$$\Delta B_y + i \Delta B_x = B_0 \sum_n (b_n + i a_n)(x+iy)^n. \qquad (5.153)$$

Here, we have placed Δ's in front of the B's to emphasize that we are concerned with field differences from the main bend field.

As suggested by our earlier treatments, identification of a resonance consists of picking out terms from the equations of motion which allow secular growth. So in writing down expressions for the rate of change of the transverse amplitudes a and b we have to identify the same term in both relations if we are talking about a two-degree-of-freedom resonance. Since the general expressions are

$$\frac{da}{dn} = \sqrt{\beta_x \beta_0} \sin \psi_x \frac{\Delta B_y l}{(B\rho)}, \tag{5.154}$$

$$\frac{db}{dn} = -\sqrt{\beta_y \beta_0} \sin \psi_y \frac{\Delta B_x l}{(B\rho)} \tag{5.155}$$

due to a single field error $(\Delta B_x, \Delta B_y)$ over a length l, then the terms we are looking for are those in which $\sin \psi_x \, \Delta B_y$ and $\sin \psi_y \, \Delta B_x$ have the same harmonic content. Take for example a resonance excited by a multipole field of order m. We look for a term in Δa which has powers $x^{m-k} y^k$; because $\sin \psi_x$ multiplies this term, it will drive a resonance which satisfies the relation $(m - k + 1)\nu_x \pm k\nu_y =$ integer. The corresponding term in Δb must contain $x^{m-k+1} y^{k-1}$, since it is multiplied by $\sin \psi_y$. Hence, the ratio of these two terms will be

$$\frac{da}{db} = \pm \frac{\binom{m}{k} a^{m-k} b^k}{\binom{m}{k-1} a^{m-k+1} b^{k-1}}, \tag{5.156}$$

where we have made use of the binomial expansion

$$(x + iy)^m = \sum_{k=0}^{m} \binom{m}{k} x^{m-k} (iy)^k, \tag{5.157}$$

$$\binom{m}{k} = \frac{m!}{k!(m-k)!}. \tag{5.158}$$

The choice of the sign of the expression depends upon whether we are considering a sum or difference resonance. Then

$$\frac{da}{db} = \pm \frac{m - k + 1}{k} \frac{b}{a}, \tag{5.159}$$

or, integrating, we find

$$\frac{a^2}{m - k + 1} \mp \frac{b^2}{k} = \text{constant}. \tag{5.160}$$

The positive sign goes with the difference resonance, and the negative sign with the sum resonance. This is the relationship we wished to obtain.

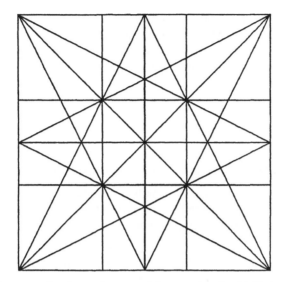

Figure 5.7. Resonance lines in a unit square of the tune plane for third integer and below.

So indeed tune relationships of the form $M\nu_x + N\nu_y = P$ can lead to resonant behavior, and in particular, when M and N have the same sign, the motion may be unbounded. It is conventional to plot resonance lines in the tune plane. In Figure 5.7 we plot all resonance lines where M and N can each take on the values 0, ± 1, ± 2, and ± 3 and $|M| + |N| \leq 3$ in a unit square in the tune plane. That is, this shows all the resonance lines up to the

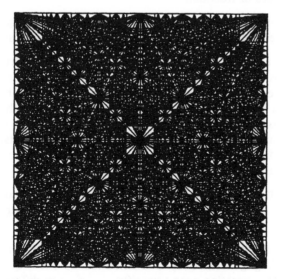

Figure 5.8. Resonance lines in the tune plane for ninth integer and below.

thirds. As we include more resonances, the tune plane rapidly becomes cluttered. Figure 5.8, which includes resonance lines up to and including ninths, rather resembles an oriental carpet. Actually, Figure 5.8 is a more realistic approximation to the resonance lines that are to be avoided in a hadron collider.

PROBLEMS

1. Derive Equation 5.1 for the magnetic field of a rotated quadrupole.

2. Derive the 4×4 matrix for motion through a solenoid magnet. The matrix can be represented as the product of three matrices: one for the body and two representing the ends.

3. Imagine a synchrotron made up of FODO cells in which all of the quadrupoles are rotated by $45°$ to become skew quadrupoles in the bend plane. Discuss the linear dynamics of this synchrotron.

4. Find the eigenvectors corresponding to the eigenfrequencies of the coupled simple harmonic oscillator problem described by Equations 5.5 and 5.6.

5. Reduce Equations 5.49–5.52 to Equations 5.55 and 5.56.

6. Consider the coupled simple harmonic oscillator problem in the text. Suppose $k_1 = k_2 = k_0$ and $k = 0.1k_0$. If $x = x_0$, $x' = y = y' = 0$ are the initial conditions, plot the ensuing motion and exhibit the beat pattern in both degrees of freedom.

7. Verify that Equation 5.106 gives Equation 5.70 when applied to the case of a single skew quadrupole in an otherwise uncoupled synchrotron.

8. Consider a large synchrotron. Suppose two skew quadrupoles of equal and opposite strength are inserted at two locations where $\beta_x = \beta_y = \beta_0$ and where $\Delta\psi_x = \Delta\psi_y = 2\pi$ between them. Show that the transverse motion in between the two skew quadrupoles is "locally" coupled, but that motion outside the pair is "globally" decoupled; to do so, compute the one-turn matrices for points just within and just outside the pair of skew quadrupoles and compare. Compute the eigenvalues of both matrices and verify that they are equal.

9. Show that the symplectic condition gives $n(2n - 1)$ independent relationships among the $2n \times 2n$ elements of an n-degree-of-freedom matrix.

10. Complete the discussion of Section 5.1.2 by generalizing the treatment to the coupling generated by a distribution of skew quadrupoles in a synchrotron operating near a specific $\nu_x - \nu_y = m$ difference resonance.

Show that two families of skew quadrupole correctors are necessary to compensate the resonance driving terms.

11. A skew quadrupole field in the presence of nonzero horizontal dispersion will generate dispersion in the vertical degree of freedom. Show that the rms spurious vertical dispersion in a large synchrotron due to random skew quadrupole fields with rms value q_{rms} would be

$$(D_y)_{rms} = \frac{q_{rms}\sqrt{\beta_0\langle\beta D_x^2\rangle}}{2|\sin \pi \nu|}\sqrt{\frac{N}{2}},$$

where the averages are taken over the N sources of skew quadrupole fields. Estimate the vertical dispersion for a 20-TeV-scale synchrotron made up of 400 FODO cells with 90 m lens spacing and 89° cell phase advance if the sources are due to random rotational misalignments of the main quadrupoles with $\phi_{rms} = 1$ mrad.

12. With the aid of a graphics terminal, generate plots of the vertical amplitude b vs. the horizontal amplitude a of a particle circulating a synchrotron with a single sextupole present. For various choices of horizontal and vertical tunes and initial conditions, verify that the coupled motion is stable near a difference resonance but can become unstable near a sum resonance.

Intensity Dependent Effects

Thus far, we have considered only the motion of a single particle, or of a beam of noninteracting particles, in the presence of external forces. Many of the interesting phenomena in accelerator physics involve the dynamic interplay between the beam and its surroundings through the electromagnetic fields initiated by the beam. Frequently—in fact, usually—the consequences are not benign, and there is a catalog of beam instabilities analogous to the more familiar instabilities of hydrodynamics.

In this chapter we introduce such effects for which the intensity of the beam is important. We begin with a discussion of space charge and its effects on betatron oscillation tunes. This will lead us to the introduction of coherent instabilities, in particular the so-called negative mass instability. The particular field used to motivate the treatment of the negative mass instability will be the longitudinal space charge field of the beam. Discussion of a general field is made possible through the introduction of impedance.

Having introduced impedance, we digress briefly with a section on the fields trailing a charge—the wake fields—and their relation with impedance. We return to the main theme, with macroparticle models of coherent instabilities that provide a simpler model for such effects as beam breakup in linacs and the head-tail effect.

A thorough treatment of coherent instabilities requires that we follow the evolution of the particle distribution function. We derive the Vlasov equation, obtain the dispersion relation for longitudinal stability of a coasting beam, and apply the result to the negative mass instability of a beam with momentum spread. We conclude with a brief general heuristic account of Landau damping.

6.1 SPACE CHARGE

Throughout most of this chapter we treat the particles constituting the beam as a continuum described by a space charge density distribution $\rho(\vec{r})$. That is, we ignore any statistical effects due to the fact that the beam is actually composed of many individual particles, and examine the motion of a single test particle under the influence of the surrounding "space charge." We should note in passing that the bulk of the present intense activity in this field lies in the low energy, high current regime where space charge forces are pervasive design considerations. For the domain of high energy accelerators it is usually possible to treat space charge effects as perturbations to the single particle motion, but their inclusion in the design process remains essential.

We will investigate transverse and longitudinal space charge effects separately. In the first case we will mainly look at examples of tune shift, essentially static field effects. Secondly, we will examine an example of a coherent instability brought on, in its simplest form, by longitudinal space charge fields.

6.1.1 The Transverse Space Charge Force

Here we begin by studying the static effects of space charge forces in unneutralized beams. We start this discussion by writing down the fields in some simple cases. First, consider a uniform cylindrical beam. If there are N particles per unit length in a beam of radius a, then, outside the beam, that is, for $r > a$, the electric and magnetic fields are

$$E = \frac{eN}{2\pi\epsilon_0 r}, \tag{6.1}$$

$$B = \frac{eNv}{2\pi\epsilon_0 rc^2}, \tag{6.2}$$

where v is the speed of the particles. Because the Coulomb repulsion is in the opposite direction to the magnetic attraction, the net outward force is

$$F = \frac{e^2 N}{2\pi\epsilon_0 r\gamma^2}, \tag{6.3}$$

where γ is the Lorentz factor. On the other hand, inside the beam, that is, for $r < a$, the fields are

$$E = \frac{eN}{2\pi\epsilon_0 a^2} r,$$ (6.4)

$$B = \frac{eNv}{2\pi\epsilon_0 a^2 c^2} r,$$ (6.5)

from which

$$F = \frac{e^2 N}{2\pi\epsilon_0 a^2 \gamma^2} r.$$ (6.6)

Note that if the beams were partially neutralized, the space charge force could be much larger. Within the beam, the linear dependence of the force on the transverse position is reminiscent of a defocusing lens.

As another example, we next consider the case of a beam which is Gaussian in both transverse coordinates with standard deviations $\sigma_x = \sigma_y \equiv \sigma$; we still assume the density is independent of the longitudinal coordinate. In this case, the fields may be written as

$$E = \frac{eN}{2\pi\epsilon_0 r} (1 - e^{-r^2/2\sigma^2}),$$ (6.7)

$$B = \frac{eNv}{2\pi\epsilon_0 rc^2} (1 - e^{-r^2/2\sigma^2}),$$ (6.8)

and the force is given by

$$F = \frac{e^2 N}{2\pi\epsilon_0 \gamma^2 r} (1 - e^{-r^2/2\sigma^2}).$$ (6.9)

Obviously, for r large compared with σ, the force varies inversely with r as one would expect. On the other hand, the force at values of r small compared with σ is

$$F = \frac{e^2 N}{4\pi\epsilon_0 \gamma^2 \sigma^2} r$$ (6.10)

and so varies linearly with transverse displacement. This suggests that particles with small oscillation amplitudes will experience a force similar to the focusing forces of beam optics considered earlier, while particles with larger amplitudes will see less of this effect.

Field distributions can be written down for more complicated beam profiles, but the subsequent treatment becomes more complex, whereas the above simple cases will illustrate the physics.

6.1.2 Equation of Motion in the Presence of Space Charge

Our previous development of the equation of motion permits the inclusion of the space charge force:

$$x'' + K(s)x = \frac{1}{\gamma m v^2} \times (\text{space charge force}). \tag{6.11}$$

Putting in the space charge force for a uniformly charged round beam, we have

$$x'' + \left[K(s) - \frac{2r_0 N}{(v/c)^2 \gamma^3 a^2} \right] x = 0, \tag{6.12}$$

where r_0 is the *classical radius* of the particle,

$$r_0 = \frac{e^2}{4\pi\epsilon_0 mc^2}. \tag{6.13}$$

(For the proton, $r_0 = 1.53 \times 10^{-18}$ m, and for the electron, $r_0 = 2.82 \times 10^{-15}$ m.) So, in a circular accelerator, the reduction in focusing strength will lead to a shift in the betatron oscillation tune. At low energies, however, the space charge force can vitiate, or indeed overcome, the external focusing force. As an example of the latter circumstance, we consider a focusing system with constant K. For low energy, where $\gamma \to 1$, the focusing force is exactly balanced by the space charge force when

$$\frac{2r_0 N}{(v/c)^2 a^2} = K \tag{6.14}$$

or, in terms of the beam current $I = eNv$,

$$I = \frac{Kec(v/c)^3 a^2}{2r_0}$$

$$= \frac{4\pi\epsilon_0 Ta^2 B'}{m}, \tag{6.15}$$

where T is the kinetic energy of the particle and the focusing force has been expressed in terms of the gradient B' of the external magnetic field.

For example, suppose we have a proton beam of 1 cm radius propagating through a field gradient of 1 T/m. Then, from the above, the ratio of the beam current to the kinetic energy must be below 1 A/MeV to ensure focusing.

6.1.3 Incoherent Tune Shift

Now let us look in more detail at the tune shift in a circular accelerator. From our discussion in Chapter 3, the change in the betatron oscillation tune due to a distribution of gradient errors is

$$\Delta\nu = \frac{1}{4\pi}\oint\frac{\beta(s)\Delta B'(s)}{(B\rho)}\,ds \rightarrow \frac{1}{4\pi}\oint\beta(s)\Delta K\,ds. \tag{6.16}$$

For our case, ΔK is

$$\Delta K = \frac{\Delta B'}{(B\rho)} \rightarrow \frac{F'}{ev(B\rho)} = \frac{F'}{pv} \tag{6.17}$$

where use has been made of the form of the Lorentz force due to a magnetic field to replace $\Delta B'$ in our formula with the gradient of a force in general. So, for a round Gaussian beam, and for small displacement compared with σ,

$$\Delta\nu = \oint\frac{1}{4\pi}\beta(s)\frac{F'(s)\,ds}{pv}$$

$$= \frac{1}{4\pi}\frac{1}{pv}\frac{e^2N}{4\pi\epsilon_0\gamma^2}\oint\frac{\beta(s)}{\sigma^2(s)}\,ds$$

$$= \frac{e^2}{4\pi\epsilon_0 mc^2}\frac{N}{(v/c)^2\gamma^3}\frac{1}{4}\oint\frac{\beta}{\pi\sigma^2}\,ds$$

$$= \frac{r_0 N}{4(v/c)\gamma^2}\oint\frac{ds}{\epsilon_N}$$

$$= \frac{\pi N r_0 R}{2\epsilon_N(v/c)\gamma^2}. \tag{6.18}$$

where R is the average radius of the accelerator, and we have used $\epsilon_N = \pi\sigma^2(\gamma v/c)/\beta$ as our definition of normalized emittance. The tune is decreased due to the defocusing character of the space charge force.

As a numerical example, consider an unbunched beam entering the Fermilab booster synchrotron. Here $R = 75$ m, the injection kinetic energy is 200 MeV, the normalized emittance of the beam at entry is π mm mrad, and the number of particles per meter is 6×10^9. The tune shift for a particle undergoing infinitesimal transverse oscillations at the center of the beam is then 0.4. It is no wonder that emittance dilution and particle loss occur under these circumstances. The remarkable thing is that tune shifts of a significant fraction of unity can be sustained within the beam. The cures for beam loss and emittance dilution due to space charge are a higher injection energy and a smaller ring.

The tune shifts for particles having larger oscillation amplitudes are smaller than those for particles at the center of the beam. Thus the beam particles span a range of tunes, with the outer particles scarcely displaced from the single particle case.

Frequently, the tune is measured by observing a coherent motion of the beam centroid. The tune so measured is not necessarily any of the values within the span of the preceding paragraph. The tune shift that we calculated above is often called an incoherent tune shift for this reason.

6.1.4 The Beam-Beam Tune Shift

A similar incoherent tune shift occurs in a colliding beam accelerator. Each time the beams cross each other, the particles in one beam feel the electric and magnetic forces due to the particles in the other beam. Consider the case of two intersecting beams of particles with like charges. Because the velocity of the "test particle" in one beam is in the opposite direction of the velocity of the oncoming beam, the electric and magnetic forces do not cancel as in the previous section, but rather add, creating a net defocusing force. For particles undergoing infinitesimal betatron oscillations in a highly relativistic Gaussian beam, the net force would be

$$F = \frac{e^2 N}{2\pi\epsilon_0 \sigma^2} r,$$
(6.19)

and the tune shift experienced by the particle would be

$$\Delta\nu = \frac{1}{4\pi} \frac{1}{pc} \frac{e^2}{2\pi\epsilon_0} \oint \frac{N\beta(s)}{\sigma^2(s)} ds$$

$$= \frac{r_0}{2\epsilon_N} \times \frac{1}{2} \int N \, ds.$$
(6.20)

In this case, the force is felt only over the time that the two bunches are colliding. Only half of the integral over the presumed symmetric distribution is necessary, because the two beams are traveling in opposite directions. As soon as the "test particle" has traveled half the bunch length, the oncoming bunch has gone past. In terms of the total number of particles in a bunch, n, the beam-beam tune shift per collision is then

$$\Delta \nu = \frac{n r_0}{4 \epsilon_N}. \tag{6.21}$$

In contrast to the somewhat larger space charge tune shift found in the previous section for a lower energy synchrotron, the beam-beam tune shift is typically rather small. For example, consider the shift for the Tevatron collider. Here, the typical numbers are $n = 6 \times 10^{10}$, $\epsilon_N = 3\pi$ mm mr. For one crossing, $\Delta \nu = 0.0025$. Up to this writing, the collider has operated with six bunches of protons and six bunches of antiprotons. This generates 12 interactions per revolution, which implies a possible tune shift of 0.03. The shift is positive due to the opposite charges involved. Unlike the case of the Booster synchrotron, where the injected beam remains in the presence of the space charge force for only a few milliseconds, the particles in the collider are stored for many hours. In view of the long-term sensitivity to resonances, one would expect that the allowed region in the tune diagram occupied by the beam particles would be quite limited. This is indeed the case.

6.1.5 Image Charge and Image Current Effects

So far we have ignored the interaction of the beam with its surroundings. Typically, a beam travels within a beam pipe made of some conducting material. So, in general, the electric field distribution will be influenced by the conducting boundary. Also, within a magnet, the magnetic field distribution will be shaped in part by the presence of magnetic materials. We take up the problem of static effects due to electric fields.

The method of images is well suited to problems of this type. Take, for example, the electric field distribution of a line charge near a perfectly conducting plane. The field lines are perpendicular to the plane everywhere, and the resulting distribution is the same as if the conducting plane were replaced by a second line distribution of opposite charge equidistant from the boundary, as shown in Figure 6.1. Therefore, the force on the line charge due to the conducting plane is the same as the force calculated due to the image.

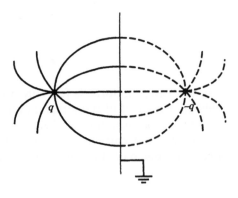

Figure 6.1. A charge near a conducting plane and its image.

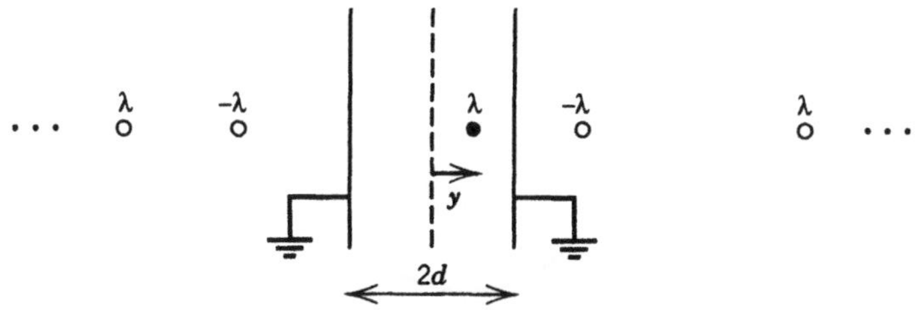

Figure 6.2. Line charge of density λ located within a rectangular beam pipe, approximated by two perfectly conducting parallel plates, and the associated images.

Of course, in a real accelerator, the beam chamber is closed. One common geometry is a chamber with rectangular cross section, where one side of the rectangle is much larger than the other. We can approximate this geometry by two parallel planes. As one would expect, in this case there are an infinite number of images. Suppose that we have a line charge with line density λ at a distance y from the center of the chamber as shown in Figure 6.2. Then at some field point y', the field due to the images alone is

$$E = \sum_i \frac{\lambda_i}{2\pi\epsilon_0 r_i} \tag{6.22}$$

$$= \frac{\lambda}{2\pi\epsilon_0}\left[\frac{1}{2d-y-y'} - \frac{1}{2d+y+y'} + \frac{1}{4d-y+y'} - \frac{1}{4d+y-y'} + \cdots\right]$$

$$= \frac{\lambda}{2\pi\epsilon_0}\left[\frac{2(y+y')}{4d^2-(y+y')^2} + \frac{2(y-y')}{16d^2-(y-y')^2} + \frac{2(y+y')}{36d^2-(y+y')^2} + \cdots\right]$$

$$\approx \frac{\lambda}{4\pi\epsilon_0 d^2}\left[\frac{y+y'}{1} + \frac{y-y'}{4} + \frac{y+y'}{9} + \frac{y-y'}{16} + \cdots\right]$$

$$= \frac{\lambda}{4\pi\epsilon_0 d^2}\left[\left(1 + \frac{1}{4} + \frac{1}{9} + \cdots\right)y + \left(1 - \frac{1}{4} + \frac{1}{9} - \cdots\right)y'\right],$$

$$= \frac{\lambda}{4\pi\epsilon_0 d^2}\left(\frac{\pi^2}{6}y + \frac{\pi^2}{12}y'\right)$$

$$= \frac{\pi\lambda}{24\epsilon_0 d^2}(y + \tfrac{1}{2}y'), \tag{6.23}$$

where we have assumed that y and y' are much less than d.

Note that the field gradient, $\partial E/\partial y'$, is independent of position y' and independent of the charge distribution. Thus the tune shift experienced by each particle in the beam is the same. This can be called a *coherent* tune shift, in contrast to the incoherent tune shift treated above.

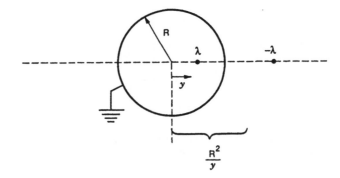

Figure 6.3. Line charge of density λ located within a round beam pipe, and the associated image.

Another simple and common geometry is that of a cylindrical beam pipe. In contrast to the parallel plate case, if a line charge is located at the center of the beam pipe, there will be no image field. But if the line charge is displaced a distance y from the center, an image line charge will appear at a distance R^2/y from the origin as shown in Figure 6.3.

Now at the field point y' the field gradient due to the image charge will be

$$\frac{\partial E}{\partial y'} = \frac{\lambda}{2\pi\epsilon_0} \frac{y^2}{R^4},$$

$$(6.24)$$

where we assume y and y' are small compared with R. So the particles of a small beam will experience a coherent tune shift if there are, for example, orbit distortions or other regions in which the beam strays from the center of the vacuum chamber.

In the above, we have used electrostatic images as examples. Magnetic images can be treated in much the same way. For vacuum chambers composed of good conductors, the electric image is the simpler because the electric field lines terminate on the chamber walls and the charges that are treated in the image method flow rapidly to preserve this boundary condition. Magnetic field lines, however, penetrate the chamber wall, and the treatment of the magnetic images must include not only the external magnetic environment, but also time constants associated with field penetration through conducting materials.

6.2 THE NEGATIVE MASS INSTABILITY

Up to now we have assumed a charge density which is independent of the longitudinal coordinate. The symmetry of this situation implies that there is no longitudinal space charge force. The first coherent instability we will examine, the negative mass instability, is driven by longitudinal space charge forces that may arise as soon as this limitation is removed.

The origin of this instability can be seen as follows. Suppose two particles in a synchrotron are traveling close together, one behind the other. Ignore for the present any fields due to their environment. The two charges will repel each other. The charge in front will gain energy and the charge behind will lose energy. Above transition, the orbit period of the first charge will increase, and that of the second will decrease. So the charges will move closer together in the longitudinal coordinate. The circumstance that a repulsive force leads to the particles approaching each other accounts for the name "negative mass."

We will discuss this phenomenon in two stages. First, we will calculate the space charge force for an unbunched beam and, as a development of the argument in the preceding paragraph, indicate how a perturbation in the charge density can grow above transition. Second, we will present a quantitative treatment for a beam without momentum spread traveling in a synchrotron with a general longitudinal impedance.[1]

6.2.1 The Longitudinal Space Charge Field

Let's suppose that the beam has a linear charge density λ that uniformly populates a cylinder of radius a. Then, provided that the derivative of λ with respect to the longitudinal coordinate s is sufficiently small, the fields from Gauss's and Ampere's laws are

$$E_r = \frac{\lambda}{2\pi\epsilon_0}\frac{1}{r}, \qquad B_\phi = \frac{\mu_0\lambda v}{2\pi}\frac{1}{r}, \qquad r \geq a; \qquad (6.25)$$

$$E_r = \frac{\lambda}{2\pi\epsilon_0}\frac{r}{a^2}, \qquad B_\phi = \frac{\mu_0\lambda v}{2\pi}\frac{r}{a^2}, \qquad r \leq a. \qquad (6.26)$$

We now find the electric field along the beam axis, as indicated in Figure 6.4. Using Faraday's law

$$\oint \vec{E} \cdot d\vec{l} = -\frac{\partial}{\partial t}\int \vec{B} \cdot d\vec{A}, \qquad (6.27)$$

---

[1]Adapted from A. Hofmann, "Single-Beam Collective Phenomena—Longitudinal," *Theoretical Aspects of the Behaviour of Beams in Accelerators and Storage Rings* (Proc. First Course of the International School of Particle Accelerators of the "Ettore Majorana" Centre for Scientific Culture, Erice, 1976), CERN 77-13, 1977.

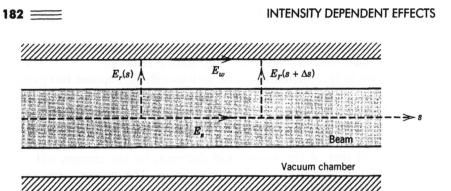

Figure 6.4. Beam passing through cylindrical beam pipe; the charge density may depend upon the longitudinal coordinate.

the left hand side is

$$(E_s - E_w)\Delta s + \left[\int_0^a \frac{r}{a^2}\, dr + \int_a^b \frac{1}{r}\, dr \right] \frac{\lambda(s + \Delta s)}{2\pi\epsilon_0}$$

$$- \left[\int_0^a \frac{r}{a^2}\, dr + \int_a^b \frac{1}{r}\, dr \right] \frac{\lambda(s)}{2\pi\epsilon_0}$$

$$= \left[(E_s - E_w) + \frac{\lambda'}{4\pi\epsilon_0}\left(1 + 2\ln\frac{b}{a} \right) \right]\Delta s, \tag{6.28}$$

where $\lambda' \equiv \partial\lambda/\partial s$. The right hand side of Equation 6.27 is

$$-\Delta s\left[\frac{1}{2} + \ln\frac{b}{a} \right]\frac{\mu_0 \dot\lambda v}{2\pi}, \tag{6.29}$$

with $\dot\lambda \equiv \partial\lambda/\partial t$. Note that the rate of change of λ with time, as observed at a particular location in the ring, is opposite in sign to the rate of change of λ with respect to s; that is,

$$\dot\lambda = -\lambda'\frac{ds}{dt} = -\lambda'v^*, \tag{6.30}$$

where v^* is the phase velocity of the wave in the charge density. The right hand side of Equation 6.27 becomes

$$\Delta s\left[\frac{\lambda'}{4\pi\epsilon_0}\left(1 + 2\ln\frac{b}{a} \right)\left(\frac{v}{c} \right)^2 \right], \tag{6.31}$$

where we assume $v^* \approx v$.

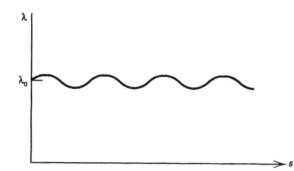

Figure 6.5. Perturbation in the line charge density λ as a function of s.

Putting the two sides together, we get

$$E_s - E_w = -\frac{\lambda'}{4\pi\epsilon_0}\left(1 + 2\ln\frac{b}{a}\right)\left[1 - \left(\frac{v}{c}\right)^2\right] \equiv -\frac{\lambda' g_0}{4\pi\epsilon_0\gamma^2}. \quad (6.32)$$

In the case of a perfectly conducting wall, E_w will vanish. In this case, we have the pure space charge field

$$E_s = -\frac{g_0\lambda'}{4\pi\epsilon_0\gamma^2}. \quad (6.33)$$

Now suppose there is a small disturbance in the charge density as that sketched in Figure 6.5. In the regions where $\lambda' > 0$ the space charge field will be negative. For $\lambda' < 0$, the field is positive. So, below the transition energy, particles in the regions where $\lambda' < 0$ will speed up and increase their revolution frequency. They will move toward the trough in the wave. Similarly, particles in the regions where $\lambda' > 0$ will slow down and again fill in the trough. The disturbance is damped. But above transition, the reverse will be the case, and the disturbance will grow.

6.2.2 Perturbation of the Line Density

We now proceed with a standard instability treatment. We assume that there is a small perturbation superimposed upon the static charge distribution and look for the circumstance under which this perturbation will grow.

Suppose the line density is perturbed according to

$$\lambda(\theta,t) = \lambda_0 + \lambda_1 e^{i(\Omega t - n\theta)}, \quad (6.34)$$

where $0 < \theta < 2\pi$ is an angular coordinate describing the azimuthal location along the unbunched beam ($\theta = s/R$). The *mode number* is denoted by n, and a general perturbation could be described by a superposition of such modes. The angular frequency of the mode is Ω; observing the disturbance at one location on the perimeter of the ring, one would see the beam vibrate at frequency Ω, while the wave propagates at angular frequency $\omega = \Omega/n$. From the continuity equation,

$$\frac{\partial \rho}{\partial t} + \nabla \cdot \vec{j} = 0, \tag{6.35}$$

λ must satisfy

$$\frac{\partial \lambda}{\partial t} + \frac{\partial}{\partial \theta}(\lambda \omega) = 0, \tag{6.36}$$

where ω is the angular frequency of the distribution. If we write ω as

$$\omega(\theta, t) = \omega_0 + \omega_1 e^{i(\Omega t - n\theta)}, \tag{6.37}$$

then the continuity equation yields

$$\omega_1 = (\Omega - n\omega_0)\frac{\lambda_1}{n\lambda_0}, \tag{6.38}$$

where second order terms have been neglected. The local beam current $I(\theta, t)$ may be written as

$$I = R\omega\lambda = I_0 + I_1 e^{i(\Omega t - n\theta)}$$
$$= R\omega_0\lambda_0 + R(\omega_0\lambda_1 + \omega_1\lambda_0)e^{i(\Omega t - n\theta)}, \tag{6.39}$$

from which we identify

$$I_0 = R\omega_0\lambda_0, \qquad I_1 = \frac{\Omega}{n}R\lambda_1. \tag{6.40}$$

We now consider an individual particle in the distribution. The rate of change of the particle's angular frequency is

$$\left(\frac{d\omega}{dt}\right)_p = \frac{\partial \omega}{\partial t} + \frac{\partial \omega}{\partial \theta}\left(\frac{d\theta}{dt}\right)_p. \tag{6.41}$$

The angular velocity of the particle is, to first order, $(d\theta/dt)_p = \omega_0$. Also, we learned in Chapter 2 that the revolution frequency of a particle is a function only of the particle's energy. That is,

$$\left(\frac{d\omega}{dt}\right)_p = \left(\frac{d\omega}{dE}\right)_p \frac{dE}{dt} = -\frac{\eta\omega_0}{(v/c)^2 E}\frac{dE}{dt}. \tag{6.42}$$

But the rate of change of the particle's energy may also be expressed in terms of an impedance to the flow of the charges in the longitudinal direction. With this concept of a longitudinal impedance, $Z_\parallel$, the rate of change of the particle's energy may be written as

$$\frac{dE}{dt} = -e(\text{energy loss per unit charge/turn})(\text{no. of turns/sec})$$

$$= -e\left(I_1 Z_\parallel e^{i(\Omega t - n\theta)}\right)\left(\frac{\omega_0}{2\pi}\right). \tag{6.43}$$

Therefore, the angular frequency must satisfy

$$\frac{\partial\omega}{\partial t} + \frac{\partial\omega}{\partial\theta}\omega_0 = \frac{e\eta I_1 Z_\parallel \omega_0^2}{2\pi(v/c)^2 E}e^{i(\Omega t - n\theta)}. \tag{6.44}$$

Substituting the original expression for ω into the above equation, we get

$$i\Omega\omega_1 - (in\omega_1)\omega_0 = \frac{e\eta I_1 Z_\parallel \omega_0^2}{2\pi(v/c)^2 E}, \tag{6.45}$$

which reduces to

$$(\Omega - n\omega_0)^2 = -i\frac{e\eta I_0 Z_\parallel \omega_0^2 n}{2\pi(v/c)^2 E}, \tag{6.46}$$

where we have made use of the fact that $R\Omega\lambda_0 \approx Rn\omega_0\lambda_0 = nI_0$.

We see immediately that if the longitudinal impedance is pure imaginary ($Z_\parallel = iZ_i$), then below transition ($\eta < 0$) the oscillations are stable if $Z_i < 0$ (i.e., if $Z_\parallel$ is capacitive). However, above transition the capacitive impedance will lead to instability. To conclude this section, we need only show that the space charge impedance is capacitive. Using Equation 6.33 for the space charge field E_s, we see that the energy loss per unit charge through one revolution is

$$-E_s 2\pi R = -\left(-\frac{g_0\lambda'}{4\pi\epsilon_0\gamma^2}\right)(2\pi R) = I_1 Z_\parallel e^{i(\Omega t - n\theta)}, \tag{6.47}$$

so that the longitudinal impedance is given by

$$Z_{\parallel} = \frac{g_0 \lambda'}{4\pi\epsilon_0\gamma^2} \frac{2\pi R}{I_1 e^{i(\Omega t - n\theta)}} = \frac{g_0 \lambda'}{4\pi\epsilon_0\gamma^2} \frac{2\pi Rn}{R\Omega\lambda_1 e^{i(\Omega t - n\theta)}}. \tag{6.48}$$

But

$$\lambda' = \frac{d\lambda}{ds} = \frac{d\lambda}{d\theta}\frac{d\theta}{ds} = -in\lambda_1 e^{i(\Omega t - n\theta)}\frac{1}{R}, \tag{6.49}$$

so that the impedance becomes

$$Z_{\parallel} = -i\frac{ng_0}{2\epsilon_0 v \gamma^2}, \tag{6.50}$$

which is indeed capacitive and hence leads to instability above transition.

6.3 WAKE FIELDS AND IMPEDANCE

In the preceding section we discussed an instability which arises from the longitudinal space charge force generated by a round beam of infinite extent as it travels through a cylindrical beam pipe of constant radius. One would like to be able to describe coherent instabilities which may arise from more general beam distributions which may be traveling through more complex geometries. In a more common situation, for example one in which the beam is bunched, it is possible that particles in the tail end of the bunch feel forces which are generated by particles in the head of the bunch interacting with the environment. As was suggested by the introduction of a longitudinal impedance in the previous section, the leading particles will lose some amount of energy; the resulting electromagnetic fields, *wake fields*, can linger in the vacuum chamber to interact with the trailing particles.

The exact form of these high frequency wake fields and their response times depend heavily on the geometry of the problem as well as the materials in the vicinity of the beam. The fields are found, of course, by solving Maxwell's equations with the appropriate boundary conditions. Much of the work nowadays is done with the aid of a variety of computer codes, but for the most part, we will stay with analytical methods and simple situations. Our goal in this section is to introduce the formal definitions of wake functions and impedance. This will allow us to discuss some examples of common beam instabilities found in high energy linacs and synchrotrons using this by now standard language. Most of this section and the next has been adapted from Chao.[2]

[2]A. W. Chao, "Coherent Instabilities of a Relativistic Bunched Beam," *Physics of High Energy Particle Accelerators* (SLAC Summer School 1982), AIP Conference Proceedings No. 105, 1983.

6.3.1 Field of a Relativistic Charge in Vacuum

To begin with, recall how one obtains the electromagnetic field of a charged particle moving in a vacuum. Though this result will be of limited use in the present context, it is a basic starting point.

The fields in the rest frame of the particle are just given by Coulomb's law. In the laboratory, the fields are related to those in the rest frame (with the prime) according to

$$E_\| = E_\|',$$ (6.51)

$$E_\perp = \gamma E_\perp',$$ (6.52)

$$B_\perp = \gamma E_\perp' v/c^2,$$ (6.53)

where "$\|$" refers to the direction of motion of the particle, and "$\perp$" means radially outward for the electric field, and looping the particle trajectory for the magnetic field. Other field components vanish.

The coordinates in the rest and laboratory frames are related according to $x' = \gamma x$ and $y' = y$. So if our field point in the laboratory is characterized by r and ψ, the distance from the charge in the rest frame will be

$$r' = \sqrt{\gamma^2 x^2 + y^2}$$

$$= r\gamma \left(1 - \frac{v^2}{c^2}\sin^2\psi\right)^{1/2}.$$ (6.54)

Using the coordinate and field transformation equations gives

$$E_\| = \frac{q}{4\pi\epsilon_0 r'^2} \frac{\gamma r \cos\psi}{r'},$$ (6.55)

$$E_\perp = \frac{q}{4\pi\epsilon_0 r'^2} \frac{\gamma r \sin\psi}{r'}.$$ (6.56)

Therefore, in the laboratory the field is pointing directly away from the charge, just as it is in the rest frame. The electric field in vector notation is thus

$$\vec{E} = \frac{q}{4\pi\epsilon_0 \gamma^2} \frac{\hat{r}}{r^2} \frac{1}{\left(1 - \frac{v^2}{c^2}\sin^2\psi\right)^{3/2}}.$$ (6.57)

As v approaches c, the field lines become concentrated perpendicular to the direction of motion, within an angular region of order $1/\gamma$. In the limit of highly relativistic motion, the fields are confined within a plane perpendicular to the direction of motion. Now that we know this result, the fields may be obtained directly from Gauss's and Ampere's laws:

$$E_r = \frac{q}{2\pi\epsilon_0 r}\delta(z - ct),$$ (6.58)

$$B_\theta = \frac{q}{2\pi\epsilon_0 cr}\delta(z - ct).$$ (6.59)

In principle, provided one knows where all the charges are, the fields can be obtained by integrating the expressions above over the charge distribution. One case where this can be done is the space charge field of a beam traveling along the axis of a perfectly conducting vacuum chamber. This problem was solved in the discussion of the negative mass instability by using Gauss's and Faraday's laws. Again, the radius of the beam is a and the radius of the vacuum chamber is b. If the charge per unit length in the beam is λ, then a surface charge $-\lambda$ is located on the inner surface of the vacuum chamber. We can find the total field by summing over the contributions of rings of charge as shown in Figure 6.6. The inner ring is a charge element of the beam, and the outer is its counterpart on the beam pipe. The amount of charge in radial range dy and in the longitudinal range dz is

$$d^2q = \lambda(z)\frac{2\pi y\, dy\, dz}{\pi a^2} = \frac{2\lambda}{a^2}y\, dy\, dz.$$ (6.60)

If the linear charge density λ is independent of z, then an integration over z will yield zero for the contribution to the field. So assume that the linear charge density has a first derivative with respect to z. The relevant charge density is then

$$d^2q = \frac{2\lambda'}{a^2}y\, dy\, z\, dz.$$ (6.61)

The contribution to the z-component of the electric field from the two rings

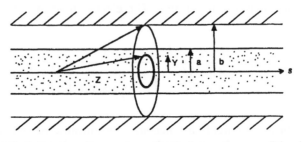

Figure 6.6. Calculation of space charge field by integrating over all the charges.

of charge is then

$$d^2E_z = -\frac{1}{4\pi\epsilon_0\gamma^2}\frac{2\lambda'}{a^2}y\,dy\left[\frac{z^2\,dz}{\left(z^2+\dfrac{y^2}{\gamma^2}\right)^{3/2}} - \frac{z^2\,dz}{\left(z^2+\dfrac{b^2}{\gamma^2}\right)^{3/2}}\right] \quad (6.62)$$

Integration over z followed by integration over y yields the same result as we obtained in the last section by a different argument:

$$E_z = -\frac{\lambda'[1+2\ln(b/a)]}{4\pi\epsilon_0\gamma^2}. \quad (6.63)$$

But generally one doesn't know where all the charges are; it then becomes necessary to solve the boundary value problem. The natural next step is to endow the beam pipe with a finite conductivity.

6.3.2 Wake Field for a Resistive Wall

There are two simple geometries that present themselves—a cylindrical beam pipe, and a rectangular chamber that is so wide that we may approximate it by two parallel plates. Let us take the former case, and immediately write down Maxwell's equations in cylindrical coordinates:

$$\frac{1}{r}\frac{\partial(rE_r)}{\partial r} + \frac{1}{r}\frac{\partial E_\theta}{\partial\theta} + \frac{\partial E_z}{\partial z} = \frac{\rho}{\epsilon_0}, \quad (6.64)$$

$$\frac{1}{r}\frac{\partial B_z}{\partial\theta} - \frac{\partial B_\theta}{\partial z} = \mu_0 j_r + \frac{1}{c^2}\frac{\partial E_r}{\partial t}, \quad (6.65)$$

$$\frac{\partial B_r}{\partial z} - \frac{\partial B_z}{\partial r} = \mu_0 j_\theta + \frac{1}{c^2}\frac{\partial E_\theta}{\partial t}, \quad (6.66)$$

$$\frac{1}{r}\frac{\partial(rB_\theta)}{\partial r} - \frac{1}{r}\frac{\partial B_r}{\partial\theta} = \mu_0 j_z + \frac{1}{c^2}\frac{\partial E_z}{\partial t}, \quad (6.67)$$

$$\frac{1}{r}\frac{\partial(rB_r)}{\partial r} + \frac{1}{r}\frac{\partial B_\theta}{\partial\theta} + \frac{\partial B_z}{\partial z} = 0, \quad (6.68)$$

$$\frac{1}{r}\frac{\partial E_z}{\partial\theta} - \frac{\partial E_\theta}{\partial z} = -\frac{\partial B_r}{\partial t}, \quad (6.69)$$

$$\frac{\partial E_r}{\partial z} - \frac{\partial E_z}{\partial r} = -\frac{\partial B_\theta}{\partial t}, \quad (6.70)$$

$$\frac{1}{r}\frac{\partial(rE_\theta)}{\partial r} - \frac{1}{r}\frac{\partial E_r}{\partial\theta} = -\frac{\partial B_z}{\partial t}. \quad (6.71)$$

For the case of a particle propagating down the axis of the cylindrical beam pipe, the charge and current densities are

$$\rho = \frac{q}{2\pi r}\delta(z - ct)\delta(r),\tag{6.72}$$

$$j_z = \frac{qc}{2\pi r}\delta(z - ct)\delta(r).\tag{6.73}$$

From the symmetry of the situation and our knowledge of the fields of the charge moving in a vacuum, we expect a solution with $B_r = 0$, $B_z = 0$, and $E_\theta = 0$. If we express the remaining field components in terms of their Fourier transforms according to

$$f(r, z, t) = \int_{-\infty}^{\infty} e^{ik(z-ct)}\tilde{f}\,dk,\tag{6.74}$$

where f is one of the field components, then within the beam pipe, transformation of Maxwell's equations yields

$$\frac{\partial \tilde{E}_r}{\partial r} + \frac{1}{r}\tilde{E}_r + ik\tilde{E}_z = \frac{q}{2\pi r\epsilon_0}\delta(r),\tag{6.75}$$

$$\tilde{B}_\theta = \frac{1}{c}\tilde{E}_r,\tag{6.76}$$

$$ik\tilde{E}_r - \frac{\partial \tilde{E}_z}{\partial r} = ikc\tilde{B}_\theta.\tag{6.77}$$

Only the transforms of the first, second, and seventh of the equations are shown. Four give no new information beyond that already conveyed by the symmetry argument, and the fourth equation is redundant.

Combination of the second and third of the set of three equations gives $\tilde{E}_z = A$, where A is a constant. Then, with this result, the solution of the first equation is

$$\tilde{E}_r = \frac{q}{2\pi\epsilon_0 r} - \tfrac{1}{2}ikAr.\tag{6.78}$$

Note that if $A = 0$, the result of Section 6.3.1 emerges. In order to find A in the present case, we have to determine the fields within the conducting wall and apply the boundary conditions. That is, we let $\vec{j} = \sigma\vec{E}$ in the material of the beam pipe, and we require E_z and B_θ to be continuous at the interface.

Again, we use the first, second, and seventh of Maxwell's equations. Application of the Fourier transformation yields

$$\frac{1}{r}\frac{\partial\left(r\tilde{E}_r\right)}{\partial r} + ik\tilde{E}_z = 0,\tag{6.79}$$

$$\tilde{B}_\theta = \frac{1}{c}\left(1 + i\frac{\mu_0 c\sigma}{k}\right)\tilde{E}_r,\tag{6.80}$$

$$ik\tilde{E}_r - \frac{\partial\tilde{E}_z}{\partial r} = ikc\tilde{B}_\theta,\tag{6.81}$$

from which we obtain for $\tilde{E}_z$ the equation

$$\frac{1}{r}\frac{\partial}{\partial r}\left(r\frac{\partial\tilde{E}_z}{\partial r}\right) + i\frac{k\sigma}{\epsilon_0 c}\tilde{E}_z = 0.\tag{6.82}$$

We anticipate that the fields do not penetrate far into the walls of the vacuum chamber. So in the two places that r appears in the equation above, we set $r = b$, where b is the radius of the beam pipe. Then, the solution of the equation for $\tilde{E}_z$ is of the form

$$\tilde{E}_z = Ae^{i\lambda(r-b)}.\tag{6.83}$$

Matching the fields at $r = b$ gives $\lambda^2 = ik\sigma/\epsilon_0 c$. In order that the fields in the material of the beam pipe remain finite for large r, the imaginary part of λ must be negative. Therefore, we can write

$$\lambda = \sqrt{\frac{|k|\sigma}{\epsilon_0 c}}\left(\frac{i + \text{sgn}\,k}{2}\right).\tag{6.84}$$

Note that the skin depth is $\delta_s = 1/\text{Im}\,\lambda$. To solve for A, we need to use the boundary condition on $\tilde{B}_\theta$.

Differentiating the second of our three Fourier transformed equations with respect to r and combining with the first gives an equation for $\tilde{B}_\theta$ in terms of $\tilde{E}_z$. Given our solution above for $\tilde{E}_z$, integration yields

$$\tilde{B}_\theta = -\frac{1}{c}\left(\frac{\lambda}{k} + \frac{k}{\lambda}\right)Ae^{i\lambda(r-b)}.\tag{6.85}$$

Then, matching the solutions at $r = b$ gives for A

$$A = \frac{q}{2\pi\epsilon_0 b\left[\frac{1}{2}ikb - \frac{k}{\lambda} - \frac{\lambda}{k}\right]}. \tag{6.86}$$

In order to proceed further analytically, let us make two approximations. First, assume that $|\lambda| \gg b^{-1}$. This is equivalent to saying that the pipe radius is large compared with the skin depth—generally a good approximation. Second, assume that $|kb| \ll |\lambda/k|$. This high frequency cutoff implies that we look at fields no closer than

$$|z - ct| \gg \left(\frac{\epsilon_0 b^2 c}{\sigma}\right)^{1/3}. \tag{6.87}$$

For typical parameters, this condition restricts us from looking closer than about 0.1 mm behind the charge responsible for the field.

Now, A reduces to

$$A = -\frac{qk}{2\pi\epsilon_0 b\lambda}, \tag{6.88}$$

and we can perform the inverse Fourier transforms to find the fields.[3] The results are

$$E_z = \sqrt{\frac{c}{2\pi\epsilon_0\sigma}}\,\frac{q}{b}\,\frac{1}{|z - ct|^{3/2}}[1 - H(z - ct)], \tag{6.89}$$

$$E_r = -\frac{3}{4}\sqrt{\frac{c}{2\pi\epsilon_0\sigma}}\,\frac{q}{b}\,\frac{r}{|z - ct|^{5/2}}[1 - H(z - ct)], \tag{6.90}$$

$$B_\theta = \frac{1}{c}E_r. \tag{6.91}$$

In the above, $H(s)$ is the Heaviside function,

$$H(s) = 1 \quad \text{if } s > 0,$$
$$= 0 \quad \text{if } s < 0, \tag{6.92}$$

and so the solutions satisfy the requirements of causality. The fact that $E_z > 0$ in Equation 6.89 means that the field is in the accelerating direction.

---

[3]See, for example, M. J. Lighthill, *Introduction to Fourier Analysis and Generalised Functions*, Cambridge, 1958, Table 1.

Apparently, E_z will change sign for sufficiently small $|z - ct|$, otherwise a net acceleration would take place. To demonstrate this, we consider the limit in which $|kb| \gg |\lambda/k|$. Then

$$\tilde{E}_z = -i\frac{q}{\pi\epsilon_0 b^2}\frac{1}{k}. \tag{6.92}$$

For $\tilde{E}_r$, because there are two terms, it is necessary to choose the range of interest for r. Let's look at the field at the surface of the beam pipe. Then in order to find $\tilde{E}_r$, one has to use the complete expression for A because the large terms cancel in the numerator. What remains is

$$\tilde{E}_r = \frac{q}{\pi\epsilon_0 b^2}\sqrt{\frac{\sigma}{\epsilon_0 c}}|k|^{-3/2}(i\,\text{sgn}\,k - 1). \tag{6.94}$$

Again making use of a table of Fourier transforms, the fields are

$$E_z = -\frac{2q}{\epsilon_0 b^2}(1 - H), \tag{6.95}$$

$$E_r = \frac{8q}{\epsilon_0 b^2}\sqrt{\frac{\sigma}{\epsilon_0 c}}|z - ct|^{1/2}(1 - H), \tag{6.96}$$

$$B_\theta = \frac{1}{c}E_r, \tag{6.97}$$

and one sees that, indeed, E_z changes sign close to the charge. An illustration of the fields is shown in Figure 6.7. Observe that the charges on the wall lag behind the particle, in contrast to the case for a perfectly conducting wall. At the wall, Poynting's vector is directed into the material, indicating that the particle loses energy by virtue of the finite conductivity of the beam pipe.

The solutions obtained thus far hold only for a beam traveling down the axis of a pipe of finite conductivity. One way of endowing the beam with an elementary shape is to use a charge distributed over a thin ring of radius a according to $\cos m\theta$. That is, the charge density representing a multipole of order $m > 0$ is

$$\rho = \frac{Q_m}{\pi a^{m+1}}\delta(z - ct)\delta(r - a)\cos m\theta, \tag{6.98}$$

where Q_m is the multipole coefficient. Note

$$Q_m = \int \rho r^m \cos m\theta\, r\, dr\, d\theta\, dz. \tag{6.99}$$

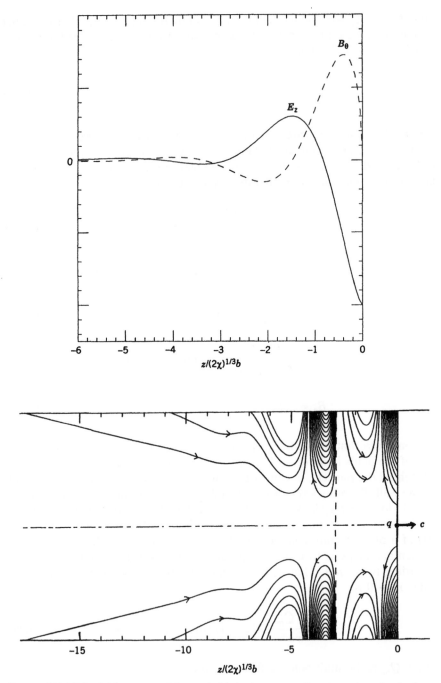

Figure 6.7. Wake fields generated by a relativistic charge traveling down the axis of a beam pipe with finite conductivity. (a) E_z and B_θ as functions of $z / (2\chi)^{1/3} b$, where $\chi = \epsilon_0 c / \sigma b$. (b) Wake electric field lines. The field line density to the left of the dashed line has been magnified by a factor of 40. (Courtesy K. Bane and A. Chao.)

For $m = 0$, the appropriate expression is

$$\rho = \frac{Q_0}{2\pi a}\delta(z - ct)\delta(r - a), \tag{6.100}$$

and hence $Q_0 = q$.

With this charge and associated current as sources, Maxwell's equations can be solved in a fashion analogous to that used for the centered charge above. Here, we will just reproduce the results from Chao within the beam pipe but not too close to the charge:

$$E_z = C_m r^m \cos m\theta \, \frac{1}{|z - ct|^{3/2}}, \tag{6.101}$$

$$E_r = -\tfrac{3}{4}C_m \frac{1}{m + 1}r^{m-1} \cos m\theta \, (r^2 + b^2)\frac{1}{|z - ct|^{5/2}}, \tag{6.102}$$

$$E_\theta = -\tfrac{3}{4}C_m \frac{1}{m + 1}r^{m-1} \sin m\theta \, (r^2 - b^2)\frac{1}{|z - ct|^{5/2}}, \tag{6.103}$$

$$B_z = -C_m r^m \sin m\theta \, \frac{1}{|z - ct|^{3/2}} \tag{6.104}$$

$$B_r = -E_\theta - 2C_m m r^{m-1} \sin m\theta \, \frac{1}{|z - ct|^{1/2}}, \tag{6.105}$$

$$B_\theta = E_r - 2C_m m r^{m-1} \cos m\theta \, \frac{1}{|z - ct|^{1/2}}, \tag{6.106}$$

where

$$C_m = \frac{Q_m}{\pi b^{2m+1}}\sqrt{\frac{2\pi c}{\epsilon_0 \sigma}} . \tag{6.107}$$

As in the preceding case, the fields vanish in front of the particle.

A couple of comments are appropriate. The radius of the ring, a, doesn't appear in these results, so for a given multipole order, the wake field is insensitive to the details of the charge distribution. For a test particle trailing the source of the wake, there will now be transverse deflecting fields at $r = 0$. And the transverse magnetic field has a long $|z - ct|^{-1/2}$ tail that will dominate at large distances.

6.3.3 Wake Functions

Instabilities develop sufficiently slowly, generally speaking, so that it is not necessary to examine the fields at each point along the trajectory; rather it is sufficient to average the fields along the trajectory. This average will depend

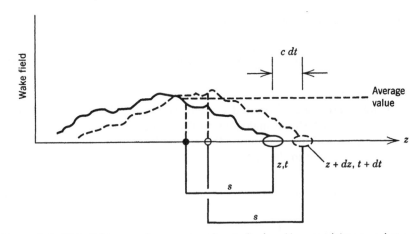

Figure 6.8. Wake fields set up by a source at longitudinal position z and time t, and at $z+dz$, $t+dt$. The repeat period of the hardware is L. A test particle trailing the source by a constant distance s will see the wake field "flutter" about an average value.

only on the separation between the source and test charges. It is, of course, the forces that we are interested in, and the equations for these forces undergo a surprising simplification as a result of this averaging.

To see this, consider a wake field which is set up by a charge distribution traveling through a structure of finite conductivity. The structure repeats itself after a length L as indicated in Figure 6.8. The wake field will have essentially the characteristics shown in Figure 6.7, but will have perturbations due to variations in the pipe structure. A test particle located a distance s behind the wake source will see an average value of the field, with time varying fluctuations. Since the fluctuations will be very fast compared to the growth rate of typical instabilities, we consider only the force on the particle due to the average fields, which will depend only upon the distance of separation

$$s \equiv ct - z \tag{6.108}$$

rather than on z and t independently. The average field components are thus functions of r, θ, and s; the time derivatives in Maxwell's equations for the average fields can be eliminated by noting that

$$\frac{df}{dt} = \left(\frac{\partial f}{\partial t}\right)_z + \left(\frac{\partial f}{\partial z}\right)_t \frac{dz}{dt} = 0$$

$$\Rightarrow \quad \frac{\partial f}{\partial t} = -c\frac{\partial f}{\partial z}, \tag{6.109}$$

where f stands for any field component, averaged over the repeat period.

Then, for example, the r-component of $\nabla \times \vec{E} = -\dot{\vec{B}}$ becomes

$$\frac{1}{r}\frac{\partial E_z}{\partial \theta} = \frac{\partial}{\partial z}(E_\theta + cB_r),\tag{6.110}$$

which, in terms of the forces, is

$$\frac{1}{r}\frac{\partial F_z}{\partial \theta} = \frac{\partial}{\partial z}F_\theta = -\frac{\partial}{\partial s}F_\theta.\tag{6.111}$$

We remind the reader that in these equations we are talking about the average fields and average forces. In an equivalent fashion, we can show that

$$ec\frac{\partial B_z}{\partial r} = \frac{\partial F_\theta}{\partial z} = \frac{1}{r}\frac{\partial F_z}{\partial \theta},\tag{6.112}$$

$$-\frac{ec}{r}\frac{\partial B_z}{\partial \theta} = \frac{\partial F_r}{\partial z} = \frac{\partial F_z}{\partial r}.\tag{6.113}$$

The solutions which satisfy these equations may be written as

$$F_r = eQ_m m r^{m-1} \cos m\theta\, W_m(s),\tag{6.114}$$

$$F_\theta = -eQ_m m r^{m-1} \sin m\theta\, W_m(s),\tag{6.115}$$

$$F_z = -eQ_m r^m \cos m\theta\, W'_m(s),\tag{6.116}$$

$$ecB_z = Q_m r^m \sin m\theta\, W'_m(s),\tag{6.117}$$

where W is a function satisfying causality and W' is the derivative of W with respect to s.

The W's are called wake functions. Often, W and W' are called the transverse and longitudinal wake functions respectively, because of their role in the solutions above. These functions provide the building blocks for the description of the forces in the time domain. Wake functions can be computed for various beam distributions and hardware geometries, and from them the forces on beam particles can be readily calculated. What is more easily obtained in the laboratory setting is the impedance presented to the beam as it passes through certain devices, through the measurement of voltage drops and currents. We have already used the impedance language in our discussion of the negative mass instability. There is a direct relationship between impedance and wake functions, which we present next.

6.3.4 Impedance

In the frequency domain, the counterpart to the wake function is the impedance. Just as there are transverse and longitudinal wake functions, there are also transverse and longitudinal impedances. In order to identify the impedance, it is natural to use a beam which is a δ-function in frequency. So, for instance, let the beam current be described by the real part of

$$I_0(z,t) = \hat{I}_0 e^{ikz - i\omega t}, \qquad (6.118)$$

where the subscript 0 implies that we are only talking about an $m = 0$ situation at present. From Equation 6.116 we can identify the longitudinal component of the electric field in terms of the wake functions W_0' due to a single point charge. Then adding up the contributions from all the charges preceding the test charge, the longitudinal electric field is

$$
\begin{aligned}
E_z(z,t) &= -\int dq \left(z, t - \frac{s'}{c} \right) W_0'(s') \\
&= -\int \frac{dq}{dt} \, dt \, W_0'(s') \\
&= -\int_0^\infty I_0 \left(z, t - \frac{s'}{c} \right) W_0'(s') \frac{ds'}{c}, \qquad (6.119)
\end{aligned}
$$

where s' is the distance from the source point to the field point. Using the above expression for I_0, the field may be rewritten as

$$E_z(z,t) = -I_0(z,t) \int_{-\infty}^\infty e^{i\omega s'/c} W_0'(s') \frac{ds'}{c}. \qquad (6.120)$$

A particle traversing some length L of the structure will then experience an energy loss due to the voltage drop $E_z L$. It is reasonable to equate this drop to the product of the current and an impedance. According to the above, we have

$$V(z,t) = -I_0(z,t) Z_0^{\parallel}(\omega) \qquad (6.121)$$

and hence may make the identification

$$\frac{Z_0^{\parallel}(\omega)}{L} = \frac{1}{c} \int_{-\infty}^\infty e^{i\omega s'/c} W_0'(s') \, ds'. \qquad (6.122)$$

The quantity $Z_0^{\parallel}$ is called the longitudinal impedance and is just the Fourier transform of the wake field. An analogous expression may be written for the $m \neq 0$ multipole moments, where W_0 is replaced by W_m.

If we treat the transverse force in an equivalent fashion, we are led to the definition of a transverse impedance. Proceeding as above, the transverse force components may be written

$$F_r = ieQ_m mr^{m-1} \cos m\theta Z_m^{\perp}(\omega), \qquad (6.123)$$

$$F_\theta = -ieQ_m mr^{m-1} \sin m\theta Z_m^{\perp}(\omega), \qquad (6.124)$$

with the transverse impedance defined as

$$\frac{Z_m^{\perp}(\omega)}{L} = \frac{1}{ic} \int_{-\infty}^{\infty} e^{i\omega s'/c} W_m(s') \, ds'. \tag{6.125}$$

The factor of i has been included to reflect the fact that the transverse force tends to be 90° out of phase with the beam current; it is just a convention. While $Z_0^{\parallel}$ has units of ohms, $Z_1^{\perp}$ (the lowest order transverse impedance) has units of ohms per meter.

For given m, there is a relationship between the longitudinal and transverse impedances. We go back to the expressions for the force components in terms of the wake functions that were written down in the preceding section. By combining the transverse components, we have

$$\nabla_{\perp} F_z = \frac{\partial}{\partial s} \vec{F}_{\perp}, \tag{6.126}$$

which is referred to as the Panofsky-Wenzel theorem. In the frequency domain, this becomes

$$Z_m^{\parallel}(\omega) = \frac{\omega}{c} Z_m^{\perp}(\omega). \tag{6.127}$$

Though we will not calculate wake functions or impedances in the remainder of this chapter, the language we have developed will be used, in a qualitative fashion, to continue our introductory treatment of coherent instabilities.

6.4 MACROPARTICLE MODELS OF COHERENT INSTABILITIES

Interestingly enough, some insights can be gained into the physics at work in a number of collective instabilities by a very simple model: the entire bunch is replaced by two macroparticles. The leading macroparticle contains half of the particles of the bunch and creates the wake field that is experienced by the members of the second macroparticle. We will apply this technique to two frequently encountered instabilities in linacs and synchrotrons.

6.4.1 Beam Breakup in Linacs

The first instability we will study using the macroparticle model is that of beam breakup in a linear accelerator. Here, the beam is represented as two macroparticles, each containing $N/2$ particles, separated by a distance s, as depicted in Figure 6.9. In an electron linac, this distance will not change with time. The second macroparticle sees the transverse wake field of the first

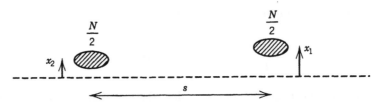

Figure 6.9. Two macroparticles, separated by a distance s, each containing half the particles of the bunch.

macroparticle which is undergoing betatron oscillations; the resulting force will drive the trailing particle's transverse oscillations.

To see this, consider the form of the transverse force due to a wake field of order $m = 1$ as presented earlier:

$$F_r = eQ_1 W_1(s). \tag{6.128}$$

Here, motion has been restricted to one degree of freedom, and

$$Q_1 = \int \rho r \cos \theta \, r \, dr \, d\theta \, dz$$

$$= \int \frac{Ne}{2} \delta(x - x_1) \delta(y) \delta(z - ct) x \, dx \, dy \, dz$$

$$= \frac{Ne}{2} x_1, \tag{6.129}$$

where x_1 is the transverse coordinate for the leading macroparticle. If the leading macroparticle is executing betatron oscillations according to

$$x_1 = \hat{x} \cos \omega_\beta t, \tag{6.130}$$

then the equation of motion of the second macroparticle becomes

$$\ddot{x}_2 + \omega_\beta^2 x_2 = \frac{Ne^2 W_1}{2m\gamma} x_1$$

$$= \frac{Ne^2 W_1}{2m\gamma} \hat{x} \cos \omega_\beta t. \tag{6.131}$$

This is the equation of a driven oscillator, where the tail of the beam is driven exactly on resonance by the head of the beam. The solution to the above differential equation is

$$x_2(t) = \hat{x}_2 \cos \omega_\beta t + \hat{x}_1 \frac{Ne^2W_1}{4\omega_\beta m\gamma} t \sin \omega_\beta t. \tag{6.132}$$

We see that on top of the free betatron oscillation, the amplitude of the motion grows linearly with time. If the two macroparticles had similar amplitudes initially, then at the end of the linac of length L, the amplitude of the oscillation in the tail will be a factor of

$$\frac{Ne^2W_1L}{4\omega_\beta m\gamma c} \tag{6.133}$$

larger than its initial value; thus intense beams can become quite distorted by the time they exit the linac.

6.4.2 The Strong Head-Tail Instability

The strong head-tail instability is basically the same as the beam breakup experienced by linac beams, but in this case the particles are undergoing synchrotron oscillations within a circular accelerator. If the synchrotron oscillations have a high enough frequency, the effect is to stabilize the beam against breakup. Above a certain intensity threshold, however, the beam can become unstable.

We begin by again considering a bunch to be composed of two macroparticles. During the first half synchrotron period, particle 1 leads particle 2 and thus particle 2 feels the effect of the wake generated by particle 1. During the second half of the synchrotron period, the roles are reversed. The equations of motion are thus

$$\left.\begin{array}{l} \ddot{x}_1 + \omega_\beta^2 x_1 = \quad 0 \\[2mm] \ddot{x}_2 + \omega_\beta^2 x_2 = \dfrac{Ne^2W_1}{2m\gamma} x_1 \end{array}\right\} \quad 0 < t < T_s/2, \tag{6.134}$$

$$\left.\begin{array}{l} \ddot{x}_1 + \omega_\beta^2 x_1 = \dfrac{Ne^2W_1}{2m\gamma} x_2 \\[2mm] \ddot{x}_2 + \omega_\beta^2 x_2 = \quad 0 \end{array}\right\} \quad T_s/2 < t < T_s. \tag{6.135}$$

It is assumed here that the wake function is a constant within the beam, but zero outside the beam.

Since we have learned how to analyze the stability of a system using 2×2 matrices, it is desirable to apply that method to the present problem. To do so, we notice that the solution to the unperturbed equation of motion,

$$x = x_0 \cos \omega_\beta t + \frac{\dot{x}_0}{\omega_\beta} \sin \omega_\beta t, \tag{6.136}$$

$$\dot{x} = \dot{x}_0 \cos \omega_\beta t - x_0 \omega_\beta \sin \omega_\beta t, \tag{6.137}$$

may be written in compact form as the phasor

$$\tilde{x}(t) = x + \frac{i}{\omega_\beta} \dot{x} = \tilde{x}(0) e^{-i\omega_\beta t}. \tag{6.138}$$

Thus, for $0 < t < T_s/2$, we have the solutions

$$x_1 = \hat{x}_1 \cos \omega_\beta t, \tag{6.139}$$

$$x_2 = \hat{x}_2 \cos \omega_\beta t + \frac{Ne^2 W_1}{4\omega_\beta m\gamma} \hat{x}_1 t \sin \omega_\beta t, \tag{6.140}$$

which can be written in phasor form as

$$\tilde{x}_2(t) = \tilde{x}_2(0) e^{-i\omega_\beta t} + i \frac{Ne^2 W_1}{4\omega_\beta m\gamma} \tilde{x}_1(0) t e^{-i\omega_\beta t} \tag{6.141}$$

if we note that the synchrotron period $T_s \gg 1/\omega_\beta$. We may then write the solution for propagation through the first half synchrotron period in matrix form as

$$\begin{pmatrix} \tilde{x}_1 \\ \tilde{x}_2 \end{pmatrix} = e^{-i\omega_\beta T_s/2} \begin{pmatrix} 1 & 0 \\ i\eta_1 & 1 \end{pmatrix} \begin{pmatrix} \tilde{x}_1(0) \\ \tilde{x}_2(0) \end{pmatrix}, \tag{6.142}$$

where

$$\eta_1 = \frac{Ne^2 W_1 T_s}{8\omega_\beta m\gamma}. \tag{6.143}$$

For the second half of the synchrotron period, the two particles simply reverse roles. Particle 1 sees the wake field created by particle 2, and particle 2 undergoes a free betatron oscillation. Thus, the solution for $T_s/2 < t < T_s$ will be

$$\begin{pmatrix} \tilde{x}_1 \\ \tilde{x}_2 \end{pmatrix} = e^{-i\omega_\beta T_s/2} \begin{pmatrix} 1 & i\eta_1 \\ 0 & 1 \end{pmatrix} \begin{pmatrix} \tilde{x}_1(T_s/2) \\ \tilde{x}_2(T_s/2) \end{pmatrix}, \tag{6.144}$$

and therefore, the motion for one complete synchrotron period is given by

$$
\begin{pmatrix} \tilde{x}_1 \\ \tilde{x}_2 \end{pmatrix} = e^{-i\omega_\beta T_s} \begin{pmatrix} 1 - \eta_1^2 & i\eta_1 \\ i\eta_1 & 1 \end{pmatrix} \begin{pmatrix} \tilde{x}_1(0) \\ \tilde{x}_2(0) \end{pmatrix}.
\tag{6.145}
$$

The motion over many synchrotron periods will be stable only if the absolute value of the trace of the matrix above is less than 2. That is, for stability, we must have

$$
\frac{Ne^2 W_1 T_s}{8\omega_\beta m\gamma} \le 2.
\tag{6.146}
$$

The above criterion tells us that for low intensity beams, the motion is stable. Once the intensity reaches a certain threshold, instability arises. One of the stabilizing factors is the synchrotron period. If the synchrotron period is small, the head and tail of the bunch switch roles more frequently and the growth found in the linac case is stabilized. The beam breakup in a linac is just the special case where the synchrotron period is infinite.

6.4.3 The Head-Tail Instability

In the above treatment of two macroparticles undergoing synchrotron oscillations, an important feature has been omitted, namely, the variation of betatron oscillation frequency with momentum. We will find that inclusion of this phenomenon places a strict criterion on the chromaticity of the accelerator.

As before, we will study the motion of two macroparticles each of charge $Ne/2$. The betatron frequency of each macroparticle now depends upon its momentum deviation $\delta \equiv \Delta p / p_0$:

$$
\begin{aligned}
\omega_\beta(\delta) &= 2\pi\nu(\delta)f(\delta) \\
&= 2\pi\nu_0 f_0 + 2\pi f_0 \xi\delta - 2\pi f_0 \nu_0 \eta\delta \\
&\approx \omega_\beta + \omega_0 \xi\delta.
\end{aligned}
\tag{6.147}
$$

We wish to use the longitudinal coordinate s as the independent variable. If we define Δt as the time interval between arrival of a particle and the arrival of the synchronous particle, then the synchrotron oscillations of the two macroparticles can be described by

$$
\Delta t_1 = \Delta \hat{t} \sin(\omega_s s/c),
\tag{6.148}
$$

$$
\Delta t_2 = -\Delta \hat{t} \sin(\omega_s s/c),
\tag{6.149}
$$

where ω_s is the synchrotron frequency. The accumulated betatron phase as a function of s is given by

$$\phi = \int \omega_\beta(\delta) \, dt = \int \frac{\omega_\beta \, ds}{c} + \omega_0 \xi \int \frac{\delta \, ds}{c} = \frac{\omega_\beta s}{c} - \frac{\omega_0 \xi \, \Delta t}{\eta}. \quad (6.150)$$

Therefore, we may write the solutions for free betatron oscillations as

$$x_1(s) = \hat{x}_1 \exp\left(-i\left[\frac{\omega_\beta s}{c} - \frac{\xi \omega_0}{\eta} \Delta \hat{t} \sin \frac{\omega_s s}{c}\right]\right), \quad (6.151)$$

$$x_2(s) = \hat{x}_2 \exp\left(-i\left[\frac{\omega_\beta s}{c} + \frac{\xi \omega_0}{\eta} \Delta \hat{t} \sin \frac{\omega_s s}{c}\right]\right). \quad (6.152)$$

For $0 < t < T_s/2$, particle 1 will obey the above equation, but particle 2 will have its equation of motion modified:

$$c^2 \frac{d^2 x_2}{ds^2} + \left[\omega_\beta + \frac{\xi \omega_0 \, \Delta \hat{t} \omega_s}{\eta} \cos\left(\frac{\omega_s s}{c}\right)\right]^2 x_2 = \frac{Ne^2 W_1}{2m\gamma} x_1, \quad (6.153)$$

where the quantity in brackets is the rate of change of the angular variable ϕ.

If we assume that x_2 is of the form given by the free oscillation, but that the amplitude $\hat{x}_2$ is allowed to change slowly with time, we may investigate the growth of the amplitude over time and look for conditions for stability. The first term in the equation above becomes

$$c^2 \frac{d}{ds} \left\{ \frac{d\hat{x}_2}{ds} \exp\left(-i\left[\frac{\omega_\beta s}{c} + \frac{\xi \omega_0}{\eta} \Delta \hat{t} \sin \frac{\omega_s s}{c}\right]\right) \right\}$$

$$\approx \left\{ c^2 \frac{d\hat{x}_2}{ds} \left(-2i\left[\frac{\omega_\beta}{c} + \frac{\xi \omega_0}{\eta} \Delta \hat{t} \frac{\omega_s}{c} \cos \frac{\omega_s s}{c}\right]\right) - c^2 \hat{x}_2 \left(\frac{d\phi}{ds}\right)^2 \right\}$$

$$\times \exp\left(-i\left[\frac{\omega_\beta s}{c} + \frac{\xi \omega_0}{\eta} \Delta \hat{t} \sin \frac{\omega_s s}{c}\right]\right), \quad (6.154)$$

assuming $\omega_s \ll \omega_\beta$ and assuming $\xi \omega_0 / \eta \, \Delta \hat{t}$ small. Substituting this back into the equation of motion and re-enforcing our approximations provides us with a relationship between the slowly varying amplitudes, namely

$$\frac{d\hat{x}_2}{ds} \approx \frac{iNe^2 W_1}{4m\gamma \omega_\beta c} \hat{x}_1 \exp\left(2i\left[\frac{\xi \omega_0}{\eta} \Delta \hat{t} \sin \frac{\omega_s s}{c}\right]\right). \quad (6.155)$$

Already having assumed $\xi\omega_0/\eta\Delta\hat{t}$ to be small, we expand the right hand side of the above and integrate to obtain

$$\hat{x}_2 = \hat{x}_2(0) + \frac{iNe^2W_1}{4m\gamma\omega_\beta c}\hat{x}_1(0)\left[s + \frac{2i\xi\omega_0\Delta\hat{t}c}{\eta\omega_s}\left(1 - \cos\frac{\omega_s s}{c}\right)\right]. \quad (6.156)$$

Thus, we may obtain a set of equations relating the amplitudes of oscillations over the first half of the synchrotron period:

$$\hat{x}_2\left(\frac{\pi c}{\omega_s}\right) = \hat{x}_2(0) + \frac{iNe^2W_1}{4m\gamma\omega_\beta c}\hat{x}_1\left(\frac{\pi c}{\omega_s} + \frac{4i\xi\omega_0\Delta\hat{t}c}{\eta\omega_s}\right)$$

$$= \hat{x}_2(0) + i\frac{\pi Ne^2W_1}{4m\gamma\omega_\beta\omega_s}\left(1 + i\frac{4\xi\omega_0\Delta\hat{t}}{\pi\eta}\right)\hat{x}_1$$

$$= \hat{x}_2(0) + i\eta_1\hat{x}_1. \quad (6.157)$$

As before, this may be written in matrix form, the matrix for one complete synchrotron period may be obtained, and a stability criterion may be found by obtaining the eigenvalues of the 2×2 matrix. If the eigenvalues are

$$\lambda_\pm = e^{i\pm\mu}, \quad (6.158)$$

then for low intensity beams ($|\eta_1| \ll 1$), $2\cos\mu \approx 2 - \eta_1^2$, or $\mu \approx \eta_1$. The imaginary part of η_1 will then give the fractional growth per synchrotron period. The growth rate in terms of time is thus given by

$$\left(\frac{1}{\tau_{\text{growth}}}\right)_\pm = \mp\frac{Ne^2W_1}{2\pi m\gamma\omega_\beta} \cdot \frac{\xi\omega_0\Delta\hat{t}}{\eta}. \quad (6.159)$$

We see that as long as the chromaticity is not equal to zero, we have growth in one of the two eigenmodes, while motion in the other eigenmode will be damped. Stability can be guaranteed only for $\xi = 0$. The more complete analysis[4] using the Vlasov equation shows that the two-particle model overestimates the growth rate of the "$-$" mode. Thus, most synchrotron storage rings are operated with slightly positive chromaticities above transition.

6.5 EVOLUTION OF THE DISTRIBUTION FUNCTION

The simplified models of the previous section are often inadequate to describe some of the finer details of beam instabilities. For instance, descriptions of beam behavior for higher mode numbers are impossible to describe

[4]Chao, *op. cit.*

with a two-particle model. One could increase the number of macroparticles, but the analysis quickly becomes complicated. Though computer tracking codes may help, it is possible to do the bookkeeping on more than a few thousand particles, whereas real beams will contain 10^{10} particles per bunch or more. To proceed, one turns to a description of the beam in terms of a continuous particle density function. The basic relationship that this function must satisfy is known as the Vlasov equation.

6.5.1 The Vlasov Equation

Consider a small region of phase space, as shown in Figure 6.10. Let the number of particles in this region, n, be given in terms of a particle density function ψ:

$$n = \psi(x,p,t)\Delta x \Delta p. \tag{6.160}$$

Each particle in the phase space is moving according to the equations of motion of the system. After an infinitesimal time interval Δt, the number of particles present in the region becomes

$$n(t + \Delta t) = n(t) + \text{flow in} - \text{flow out}. \tag{6.161}$$

Consider, for a moment, only the flow in the x-direction. The number of particles entering the box from the left is

$$\psi(x, p, t)\Delta p \dot{x}(x, p, t)\Delta t, \tag{6.162}$$

whereas the number exiting the box on the right is

$$\psi(x + \Delta x, p, t)\Delta p \dot{x}(x + \Delta x, p, t)\Delta t. \tag{6.163}$$

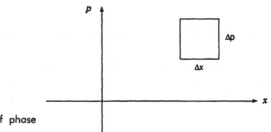

Figure 6.10. Infinitesimal area of phase space.

Therefore, the rate of change of particles in the region is given by

$$\frac{n(t + \Delta t) - n(t)}{\Delta t} = \psi(x, p, t)\Delta p\dot{x}(x, p, t)$$

$$- \psi(x + \Delta x, p, t)\Delta p\dot{x}(x + \Delta x, p, t)$$

$$= \psi(x, p, t)\Delta p\dot{x}(x, p, t) - \psi(x, p, t)\Delta p\dot{x}(x, p, t)$$

$$- \frac{\partial\psi}{\partial x}(x, p, t)\Delta x\Delta p\dot{x}$$

$$= -\frac{\partial\psi}{\partial x}(x, p, t)\Delta x\Delta p\dot{x}, \qquad (6.164)$$

or

$$\frac{\psi(t + \Delta t) - \psi(t)}{\Delta t}\Delta x\Delta p = -\frac{\partial\psi}{\partial x}(x, p, t)\Delta x\,\Delta p\dot{x}, \qquad (6.165)$$

from which

$$\frac{\partial\psi}{\partial t} + \dot{x}\frac{\partial\psi}{\partial x} = 0. \qquad (6.166)$$

Performing the same argument for both x and p yields the Vlasov equation:

$$\frac{\partial\psi}{\partial t} + \dot{x}\frac{\partial\psi}{\partial x} + \dot{p}\frac{\partial\psi}{\partial p} = 0. \qquad (6.167)$$

6.5.2 The Dispersion Relation

The solution to the Vlasov equation contains more information than we may be interested in knowing. Usually we just want to know if a situation is stable or unstable and, if unstable, what the growth rate of a perturbation would be. This determination can be expressed as a relationship between the unperturbed distribution and the perturbing forces. Expressions of this general type are called dispersion relations because they were first found in the analysis of the dispersive properties of optical materials; the name, of course, does not have this significance here.

We will use the Vlasov equation to develop a dispersion relation for particle beams with momentum spread. If $\theta \equiv s/R$ is the longitudinal coordinate and $\delta \equiv \Delta p/p$ is the conjugate momentum variable, then the Vlasov equation is

$$\frac{\partial\psi}{\partial t} + \dot{\theta}\frac{\partial\psi}{\partial\theta} + \dot{\delta}\frac{\partial\psi}{\partial\delta} = 0, \qquad (6.168)$$

where

$$\dot{\theta} = \omega, \tag{6.169}$$

$$\dot{\delta} = \frac{1}{(v/c)^2} \frac{\Delta \dot{E}}{E}$$

$$= -\frac{e\omega_0 I_1 Z_\parallel}{2\pi(v/c)^2 E} e^{i(\Omega t - n\theta)}, \tag{6.170}$$

with Equation 6.43 being used in the last step.

If the unperturbed particle distribution is independent of θ (unbunched beam), then we may write ψ as

$$\psi(\delta, \theta, t) = \psi_0(\delta) + \psi_1(\delta) e^{i(\Omega t - n\theta)}. \tag{6.171}$$

So, to first order,

$$i(\Omega - n\omega)\psi_1 - \frac{\partial \psi_0}{\partial \delta} \frac{e\omega_0 I_1 Z_\parallel}{2\pi(v/c)^2 E} = 0. \tag{6.172}$$

But

$$\frac{\partial}{\partial \delta} = -\omega_0 \eta \frac{\partial}{\partial \omega}, \tag{6.173}$$

so that

$$\psi_1 = \frac{ie\omega_0^2 I_1 Z_\parallel \eta}{2\pi(v/c)^2 E} \frac{\partial \psi_0}{\partial \omega} \frac{1}{\Omega - n\omega}. \tag{6.174}$$

If we now integrate both sides over ω and then multiply by $ev/R = e\omega_0$, we obtain

$$e\omega_0 \int \psi_1(\omega) d\omega = -e\omega_0^2 \eta \int \psi_1(\delta) d\delta = -\omega_0 \eta I_1$$

$$= i\frac{e^2 \omega_0^3 I_1 Z_\parallel \eta}{2\pi(v/c)^2 E} \int \frac{(\partial \psi_0/\partial \omega)}{\Omega - n\omega} d\omega, \tag{6.175}$$

so that we have the dispersion relation

$$1 = -i\frac{e^2 \omega_0^2 Z_\parallel}{2\pi(v/c)^2 E} \int \frac{(\partial \psi_0/\partial \omega)}{\Omega - n\omega} d\omega. \tag{6.176}$$

6.5.3 Application to the Negative Mass Instability

Let's use the above dispersion relation to investigate the negative mass instability. First, we examine the case of an unbunched beam with no momentum spread as considered earlier. In this case, the phase space density is represented by a Dirac δ-function

$$\psi_0(\delta,\theta,t) = \frac{N}{2\pi}\delta(\delta). \tag{6.177}$$

where N is the total number of particles in the distribution. Written in terms of the angular frequency,

$$\psi_0(\omega,\theta) = -\frac{N\eta\omega_0}{2\pi}\delta(\omega - \omega_0). \tag{6.178}$$

We then obtain

$$\int\frac{\partial\psi_0/\partial\omega}{\Omega - n\omega}d\omega = -\frac{N\eta\omega_0}{2\pi}\int\frac{\delta'(\omega - \omega_0)}{\Omega - n\omega}d\omega$$

$$= \frac{N\eta\omega_0}{2\pi}\frac{n}{(\Omega - n\omega)^2}, \tag{6.179}$$

which, when substituted into the dispersion relation, gives us

$$1 = -i\frac{e^2\omega_0^2 Z_\parallel}{2\pi(v/c)^2 E}\frac{N\eta\omega_0}{2\pi}\frac{n}{(\Omega - n\omega)^2}, \tag{6.180}$$

or

$$(\Omega - n\omega)^2 = -i\frac{e\eta Z_\parallel\omega_0^2 n}{2\pi(v/c)^2 E}\left(\frac{eN\omega_0}{2\pi}\right) = -i\frac{e\eta I_0 Z_\parallel\omega_0^2 n}{2\pi(v/c)^2 E}, \tag{6.181}$$

which is the same result obtained previously.

To go one step further, we look at a more realistic beam, one in which there is a distribution in momentum space. As an example, consider an unbunched beam with a Gaussian distribution in momentum. The density function is of the form

$$\psi_0(\delta,\theta) = \frac{N}{(2\pi)^{3/2}\sigma}e^{-\delta^2/2\sigma^2}, \tag{6.182}$$

where $\sigma = (\Delta p/p)_{rms}$. In terms of the angular frequency,

$$\psi_0(\omega) = \frac{N}{(2\pi)^{3/2}\sigma}e^{-(\omega-\omega_0)^2/2(\eta\omega_0\sigma)^2}. \qquad (6.183)$$

The dispersion integral becomes

$$\int_{-\infty}^{+\infty}\frac{\partial\psi_0}{\partial\omega}\frac{1}{\Omega - n\omega}d\omega = -\frac{N}{(2\pi)^{3/2}\sigma}\frac{1}{(\eta\omega_0\sigma)^2}$$

$$\times\int_{-\infty}^{+\infty}\frac{\omega - \omega_0}{\Omega - n\omega}e^{-(\omega-\omega_0)^2/2(\eta\omega_0\sigma)^2}d\omega$$

$$= \frac{N}{(2\pi)^{3/2}\sigma}\frac{1}{\eta\omega_0\sigma n}$$

$$\times\int_{-\infty}^{+\infty}\frac{\omega - \omega_0}{\omega - \Omega/n}e^{-(\omega-\omega_0)^2/2(\eta\omega_0\sigma)^2}d\left(\frac{\omega - \omega_0}{\eta\omega_0\sigma}\right)$$

$$= \frac{N}{(2\pi)^{3/2}\eta\omega_0\sigma^2 n}\int_{-\infty}^{+\infty}\frac{u}{u - u_0}e^{-u^2/2}du, \qquad (6.184)$$

where

$$u \equiv \frac{\omega - \omega_0}{\eta\omega_0\sigma}, \qquad (6.185)$$

$$u_0 \equiv \frac{\Omega - n\omega_0}{\eta\omega_0\sigma n} = \frac{\Delta\Omega}{\eta\omega_0\sigma n}. \qquad (6.186)$$

Then the dispersion relation becomes

$$1 = -i\frac{eI_0 Z_\parallel}{2\pi(v/c)^2 E\eta\sigma^2 n}I_D(u_0), \qquad (6.187)$$

where

$$I_D(u_0) \equiv \frac{1}{\sqrt{2\pi}}\int_{-\infty}^{+\infty}\frac{u}{u - u_0}e^{-u^2/2}du, \qquad (6.188)$$

and we have identified $e\omega_0 N/2\pi$ as the current I_0.

We have seen that for a beam with zero momentum spread, a capacitive impedance ($Z_\parallel = iZ_i$, $Z_i < 0$) leads to instability above transition. We now use the above dispersion relation to look for a condition that will assure stability to a beam with a Gaussian momentum distribution.

Consider the case where $\Delta\Omega$ has a negative imaginary part, corresponding to an unstable solution. The dispersion integral can be found after noting that

$$\frac{1}{u - u_0} = -i \int_0^\infty e^{i(u-u_0)\alpha} \, d\alpha, \tag{6.189}$$

if u_0 has a negative imaginary part. Then

$$\begin{aligned}
I_D(u_0) &= \frac{-i}{\sqrt{2\pi}} \int_{-\infty}^{+\infty} u e^{-u^2/2} \int_0^\infty e^{i(u-u_0)\alpha} \, d\alpha \, du \\
&= \frac{-i}{\sqrt{2\pi}} \int_0^\infty e^{-iu_0\alpha} \int_{-\infty}^{+\infty} u e^{-(u^2 - 2iu\alpha - \alpha^2)/2} e^{-\alpha^2/2} \, du \, d\alpha \\
&= \frac{-i}{\sqrt{2\pi}} \int_0^\infty e^{-iu_0\alpha} e^{-\alpha^2/2} \int_{-\infty}^{+\infty} u e^{-(u-i\alpha)^2/2} \, du \, d\alpha \\
&= -i \int_0^\infty e^{-iu_0\alpha} e^{-\alpha^2/2} i\alpha \, d\alpha \\
&= \int_0^\infty \alpha e^{-iu_0\alpha} e^{-\alpha^2/2} \, d\alpha. \tag{6.190}
\end{aligned}$$

Notice that $I_D(0) = 1$. Therefore, for Im $u_0 < 0$, $|I_D(u_0)| < 1$. So, from the dispersion relation, if in addition

$$\frac{eI_0|Z_\parallel|}{2\pi(v/c)^2 E\eta\sigma^2 n} < 1, \tag{6.191}$$

then the dispersion relation cannot be satisfied, implying that there will not be any unstable solutions. That is,

$$\sigma^2 > \frac{eI_0}{2\pi\eta(v/c)^2 E} \left|\frac{Z_\parallel}{n}\right| \tag{6.192}$$

is a sufficient condition for stability.

The general requirement

$$\left|\frac{Z_\parallel}{n}\right| < \mathscr{F}\frac{2\pi|\eta|(v/c)^2 E\sigma^2}{eI_0} \tag{6.193}$$

for stability, where $\mathscr{F}$ is a form factor of order unity which depends upon the particle distribution used, is known as the Keil-Schnell criterion.[5]

[5]E. Keil and W. Schnell, CERN Report CERN/ISR-TH-RF/69-48, 1969.

Finally, it is interesting to note that the integral $I_D(u_0)$ can be written in closed form as[6]

$$I_D(u_0) = e^{-u_0^2/4} D_{-2}(\pm i u_0), \tag{6.194}$$

where the plus sign is used for Im $u_0 < 0$, the negative sign for Im $u_0 > 0$. The function $D_{-2}(x)$ is a parabolic cylinder function and is described in Abramowitz and Stegun.[7]

6.6 LANDAU DAMPING

In the last section, we saw that there was a threshold for instability of a nondissipative system—nondissipative because in the example of space charge forces only, the impedance is reactive. For sufficiently small values of the impedance the system is stable. Buried within the mathematics of the dispersion relation is a stabilizing mechanism, which is called Landau damping. In this section we describe its origin.

Let's go back to the driven harmonic oscillator:

$$\ddot{x} + \omega_0^2 x = C \sin \omega t. \tag{6.195}$$

Consider first the situation on resonance, where $\omega = \omega_0$. For a particle starting from rest, the solution to the above differential equation is

$$x = \frac{C}{2\omega_0^2} \sin \omega_0 t - \frac{C}{2\omega_0} t \cos \omega_0 t. \tag{6.196}$$

Clearly, the envelope of the oscillation grows without bound. Off resonance,

$$x = \frac{C}{\omega_0^2 - \omega^2} \left(\sin \omega t - \frac{\omega}{\omega_0} \sin \omega_0 t \right)$$

$$= \frac{C}{(\omega_0 + \omega)\omega_0} \sin \omega_0 t$$

$$- \frac{C}{(\omega_0 + \omega)} t \left[\frac{\sin \frac{1}{2} \delta \omega \, t}{\frac{1}{2} \delta \omega \, t} \right] \cos\left(\frac{\omega + \omega_0}{2} t \right), \tag{6.197}$$

[6]I. S. Gradshteyn, and I. M. Ryzhik, *Table of Integrals, Series, and Products*, Academic Press, New York, 1980, p. 337, formula 3.462.
[7]M. Abramowitz and I. A. Stegun, *Handbook of Mathematical Functions*, Dover, New York, 1970, p. 710. In this reference, the function $D_{-2}(x)$ is denoted by $U(1.5, x)$.

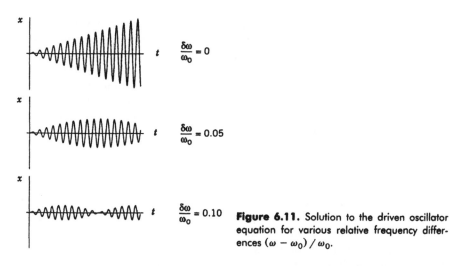

Figure 6.11. Solution to the driven oscillator equation for various relative frequency differences $(\omega - \omega_0) / \omega_0$.

where $\delta\omega \equiv \omega - \omega_0$. If $\delta\omega$ is small compared with ω_0, the two solutions are identical for t small. That is, the particle does not know initially whether or not it is on resonance. But at $t \approx 1/\delta\omega$ the solutions become distinct. The behavior is illustrated in Figure 6.11.

The circumstance that the fraction of the particles participating in secular growth diminishes with time suggests that the constant amplitude force does not produce an instability in the sense that we are looking for here. In contrast, let us consider a force that increases exponentially with time. Since we are looking for exponentially growing solutions it is reasonable to consider forces which grow exponentially as well. Suppose the equation of motion is of the form

$$\ddot{x} + \omega^2 x = Ae^{i\omega_0 t}e^{t/\tau}. \tag{6.198}$$

For $\omega = \omega_0$, the particular solution may be written as

$$x = \frac{A\tau}{2i\omega_0}e^{i\omega_0 t}e^{t/\tau} \tag{6.199}$$

for large τ. The off-resonance case $(\omega \neq \omega_0)$ has the particular solution

$$x = \frac{A}{\omega^2 - \omega_0^2 + 2i\omega_0/\tau}e^{i\omega_0 t}e^{t/\tau}. \tag{6.200}$$

The amplitude of this solution, for $\omega \approx \omega_0$, is

$$\frac{A}{2\omega_0\delta\omega + 2i\omega_0/\tau} \tag{6.201}$$

where $\delta\omega = \omega - \omega_0$. If we consider an ensemble of particles with frequencies distributed about ω_0, then those particles within a frequency range $\delta\omega \approx 2/\tau$ will participate in the growth.

In the case of a coherent instability, the driving force is not external but is due to the deviation of the beam itself. We can extend the harmonic oscillator example above to contain such a driving force. For instance, assume that the force is proportional to the average displacement. Each member of the ensemble obeys an equation of the form

$$\ddot{x} + \omega^2 x = c\langle x \rangle. \tag{6.202}$$

Suppose $\Delta\omega$ characterizes the frequency width of the distribution. At some time, the fraction of the particles participating in the growth will be $2/\Delta\omega\,\tau$. Since these particles all will be at roughly the same displacement, $\langle x \rangle \sim (2/\Delta\omega\,\tau)x$. All of these particles have essentially the same frequency, ω_0. Thus, for one of these particles, the equation of motion is

$$\ddot{x} + \left(\omega_0^2 - \frac{2}{\Delta\omega\,\tau}c\right)x = 0. \tag{6.203}$$

If the particle's amplitude is to undergo resonant growth, the frequency ω, given by

$$\omega^2 = \omega_0^2 - \frac{2}{\Delta\omega\,\tau}c, \tag{6.204}$$

must have a negative imaginary part. That is,

$$\mathrm{Im}\,\omega = \mathrm{Im}\sqrt{\omega_0^2 - \frac{2}{\Delta\omega\,\tau}c}$$

$$= \mathrm{Im}\left\{\omega_0\left(1 - \frac{c}{\Delta\omega\,\omega_0^2\tau}\right)\right\} = -\frac{\mathrm{Im}\,c}{\Delta\omega\,\omega_0\tau} < 0. \tag{6.205}$$

So for $\mathrm{Im}\,c > 0$, the oscillation amplitude grows like

$$\exp\left[\frac{(\mathrm{Im}\,c)t}{\Delta\omega\,\omega_0\tau}\right]. \tag{6.206}$$

But the argument of the exponential is just t/τ, and so for the onset of instability one must have

$$\mathrm{Im}\,c \approx \omega_0\Delta\omega. \tag{6.207}$$

Notice that the larger the frequency spread $\Delta\omega$, the larger the force coefficient must become in order to produce instability. In the preceding section, we found that the negative mass instability would be stabilized by sufficient momentum spread. Since particles of different momenta have different revolution frequencies, the conclusions of this section and of the preceding one are related. For sufficiently large frequency spread the feedback mechanism that is the potential source of instability is not strong enough to produce exponential growth. The stabilizing mechanism is called Landau damping.

PROBLEMS

1. For the round Gaussian beam, sketch as a function of r the space charge force and its derivative with respect to r. Calculate the values of r/σ for which F and dF/dr are maximum.

2. The dependence of tune on amplitude for the round Gaussian beam can be calculated analytically. In this problem, ignore the details of alternating gradient focusing and assume that a betatron oscillation in the absence of space charge is $x = a \cos\psi$, where ψ just advances linearly with azimuth, $\psi = \nu\theta$, $0 < \theta < 2\pi$.

(a) Including space charge, show that the equation of motion is

$$\frac{d^2u}{d\psi^2} + u = 4\frac{\Delta\nu}{\nu}\frac{1}{u}(1 - e^{u^2/2}),$$

where $u \equiv x/\sigma$ and $\Delta\nu$ is the magnitude of the incoherent space charge tune shift as calculated in the text.

(b) Next, assume that the solution of the equation of motion will be of the form

$$u = u_0 \cos\left[\psi + \frac{\delta(u_0)}{\nu}\psi\right],$$

where δ is the quantity that we are looking for. Use the method of phase averaging to extract this "DC" quantity. Substitute the trial solution into the equation of motion and average over ψ. Show that the amplitude dependence of tune is given by

$$\delta(u_0) = -\frac{\Delta\nu}{(u_0/2)^2}\left\{1 - e^{-(u_0/2)^2}I_0\left[(u_0/2)^2\right]\right\}$$

where I_0 is the modified Bessel function of order zero.

3. Compare the results of the phase averaging method in the preceding problem with the amplitude dependence of tune obtained by numerical integration of the equation of motion.

4. In the Fermilab booster, suppose that the injected beam is very rapidly neutralized; that is, there is quickly only a current and no charge density. Using the parameters in Section 6.1.3, calculate the tune shift for this completely neutralized beam.

5. In a proton-antiproton collider, the beam-beam tune shift parameter $\Delta\nu$ that was calculated in the text will be positive. So particles with small betatron oscillation amplitudes will have higher tunes than those with larger amplitudes, and the largest amplitude particles will have the tune prescribed by the accelerator lattice, ν. By choice of $\Delta\nu$ and ν, one can arrange that a low order resonance will be encountered at an intermediate amplitude. With the aid of a graphics terminal, develop the turn by turn mapping in one-degree-of-freedom phase space, and demonstrate the chain of four resonance islands that appear when a quarter integer resonance is within the beam. Note that the resonance does not lead to beam loss; for realistic parameters, no trajectories go to large amplitude.

6. Rewrite the expression for the luminosity of a collider found in Chapter 1 in terms of the normalized emittance (39%), and then eliminate the emittance in favor of the beam-beam tune shift parameter, $\Delta\nu_{bb}$. For the Tevatron operating at 1 TeV, with 5×10^{10} particles per bunch in each of six proton and six antiproton bunches, the tune shift parameter is near the upper limit of $\Delta\nu_{bb} = 0.007$. Calculate the luminosity under these conditions if the value of the amplitude function at the interaction point is $\beta^* = 0.5$ m. (The radius of the collider is 1 km.)

7. In electron-positron colliders, at low beam current I the luminosity varies as I^2; then there is a discontinuity in slope beyond which the luminosity varies as I. Account for this behavior, under the assumption that the beam-beam effect is a limitation.

8. In colliders designed to have a short bunch spacing, it is desired that the beams be brought into collision with a crossing angle α so as to minimize the number of head-on beam-beam collisions. The intent is that the beam-beam tune shift will be kept in bounds. However, the bunches will still act upon each other at a distance, and so we have to keep the so-called long range beam-beam tune shift in mind. Suppose that the crossing angle is such that the two Gaussian bunches of equal total charge pass each other a distance d apart, where $d \gg \sigma$.

   (a) Compute the force on a particle in one bunch due to the other bunch. Show that there is a dipole term and a quadrupole term.

   (b) Let $\Delta\nu_{HO}$ denote the beam-beam tune shift due to a single head-on collision between two bunches at the interaction point. Show that the

the long-range tune shift is

$$\Delta \nu_{LR} = \frac{2\Delta \nu_{HO}}{(d/\sigma)^2}.$$

(c) The head-on beam-beam force is defocusing in both degrees of freedom for like sign beams; what about the long range force?

(d) Suppose that the two beams collide at the interaction point, where the amplitude function has its minimum value β^*. Suppose that β^* is much smaller than the distance to the point at which long range passages take place. Show that

$$\left(\frac{d}{\sigma}\right)^2 = \frac{\pi \gamma \alpha^2 \beta^*}{\epsilon_N},$$

where

$$\epsilon_N = \left(\gamma \frac{\upsilon}{c}\right) \pi \frac{\sigma^2}{\beta},$$

and is the phase space area enclosing some 39% of the beam in one transverse degree of freedom.

(e) If the bunch spacing is S_B, show that the total long range beam-beam tune shift across the region of length L is

$$\Delta \nu_{LR} = \frac{N r_0 L}{\pi \gamma \alpha^2 \beta^* S_B}$$

$$= \Delta \nu_{HO} \left(\frac{4L \epsilon_N}{\pi \gamma \alpha^2 \beta^* S_B}\right).$$

(f) Calculate $\Delta \nu_{HO}$ and $\Delta \nu_{LR}$ for a 20 TeV collider with $\beta^* = 0.5$ m, a crossing angle, $\alpha = 0.075$ mrad, and $S_B = 5$ m, $N = 10^{10}$, $L = 180$ m.

(g) Estimate the steering error caused by the dipole component of the long range force for the parameters above.

9. It was stated in the text that if a line charge is displaced a distance y from the center of a beam pipe of radius R, an image charge equal in magnitude but opposite in sign will appear at a distance R^2/y. Prove that such is the case.

10. Estimate the coherent tune shift experienced by the particles in a bunch of 2×10^{10} protons traveling in a wide rectangular conducting vacuum chamber of height 5 cm. Assume that the bunch is 1 meter in length, the

kinetic energy is 8 GeV, the average value of the amplitude function is 50 meters, and the orbit radius is 1000 meters. These parameters correspond to the Fermilab Main Ring at injection.

11. In the resistive wake, close to the charge, obtain the next order term by taking

$$A = \frac{q}{i\pi\epsilon_0 kb^2}\left(1 + \frac{2\lambda}{ik^2b}\right).$$

Compare with the figures in the text.

12. For the resistive wake, find the wake functions W_0 and W_0'.

13. By Fourier transformation of W_0', find the longitudinal impedance $Z_\parallel$ for the resistive wake.

14. Estimate the longitudinal impedance presented to the beam passing through a bellows. One can proceed as follows. Calculate the flux "trapped" in the shaded area as though $\vec{B}$ were given by the Biot-Savart Law. If we call this flux Φ, then Faraday's law can be used to calculate the back emf on the beam. Show that the contribution of the bellows to Z/n is

$$iZ_0\left(\frac{(v/c)g}{2\pi R}\right)\ln\frac{d}{b},$$

where Z_0 is the impedance of free space.

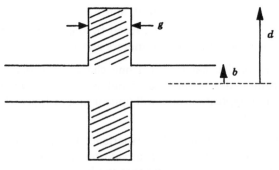

Figure 6-A.

15. Verify that in the "+" mode of the head-tail instability the two macroparticles are moving in phase, while in the "−" mode they are 180° out of phase.

16. Consider the Fermilab Main Ring at its maximum energy of 150 GeV. Assume $Z_\parallel/n = 10\ \Omega$. Suppose that 2×10^{13} protons are circulating in 1000 bunches, and that the peak instantaneous current I_0 is 10 times the average current. The smallest the mode number can be is thus $\sim 10^4$. Calculate the growth rate of the negative mass instability in this circumstance. What frequency would be detected for this motion? (For this reason the negative mass instability for a bunched beam is usually called the microwave instability.)

17. Find the intensity threshold N for the strong head-tail instability in the Tevatron at injection. Take $W_1 \approx 4 \times 10^{15}\ \mathrm{F^{-1}\,m^{-2}}$, $T_s = 12$ msec, $\nu = 19.4$, and a revolution frequency of 47.7 kHz. Is this result realistic?

18. Show that Equation 6.202 leads to the dispersion relation

$$1 = \frac{c}{2\pi} \int \frac{\rho(\omega)\, d\omega}{\omega - \Omega}$$

where $\rho(\omega)$ is the distribution function of oscillator frequencies normalized to unity.

19. Show that the dispersion integral in Problem 18 becomes

$$\int_{-\infty}^{\infty} \frac{\rho(\omega)\, d\omega}{\omega - \Omega} = \mathrm{P.V.} \int_{-\infty}^{\infty} \frac{\rho(\omega)\, d\omega}{\omega - \Omega_r} - i\pi\rho(\Omega_r)$$

for Ω_i sufficiently small, that is, at threshold.

20. Apply the threshold condition of the preceding problem to a triangular frequency distribution. Let $\rho(\omega)$ be centered at ω_0, with width $2\,\Delta\omega$ at the base, and $\Delta\omega \ll \omega_0$. At the apex of the triangle, $\rho(\omega_0) = 1/\Delta\omega$.
   (a) If $\mathrm{Re}\,\Omega = \omega_0$, show that the threshold for instability is

$$c = i\frac{2\omega_0\,\Delta\omega}{\pi}.$$

   (b) If $\mathrm{Re}\,\Omega = \omega_0 + \Delta\omega$, show that $\mathrm{Im}\,c$ can be vanishingly small and the collective motion will still be unstable.
   (c) Sketch the region of stability in $\mathrm{Im}\,c$ (y-axis) versus $\mathrm{Re}\,c$ coordinates.

21. The discussion connected with Equations 6.202 through 6.207 may be more persuasive if carried out in terms of energy. That is, compare the

power required to produce amplitude growth with the power that the feedback mechanism is able to deliver. As before, take Re $\Omega = \omega_0$, and make the same assumptions concerning the fraction of the oscillators participating in the coherent motion. Remember that because energy involves the square of displacement or velocity, you must take the real part of displacement or velocity before squaring. Show that the threshold value for Im c found by this method is essentially the same as that obtained by the argument in the text.

22. Imagine a distribution of particles of the form

$$\psi_0 = \frac{15N}{32\pi}\left[1 - \left(\frac{\delta}{\delta_{max}}\right)^2\right]^2$$

where $\delta \equiv \Delta p/p$. Use the dispersion relation given by Equation 6.176 to find the locus of points in Re $Z_\parallel$, Im $Z_\parallel$ space at the threshold of instability.

23. The extraction kinetic energy of the Fermilab booster is 8 GeV. Suppose 2×10^{12} particles at that energy are circulating in an orbit whose frequency is 6×10^5 Hz. Suppose the bunching factor is 10. The beam size is roughly one-tenth the aperture at this energy. Calculate the momentum spread necessary to stabilize the microwave instability if it originates only from space charge. For this accelerator, $\gamma_t = 5.4$. Comment on the stability threshold near transition.

CHAPTER 7

Emittance
Preservation

In our main examples of accelerator applications in Chapter 1—the luminosity of a collider and the brightness of a synchrotron light source—we commented on the importance of producing and maintaining small beam size. The beam quality aspect of small beam size is small emittance. While it was demonstrated in Chapter 3 that the properly normalized emittances are adiabatic invariants, there are, unfortunately, a variety of processes which will lead to emittance growth.

Examples of this sort include the various scattering and diffusion processes afflicting a beam. The scattering of beam particles by interactions with the residual gas in the vacuum chamber will lead to emittance growth and beam loss. Scattering among the particles of a single beam can lead to growth of the beam dimensions in all three degrees of freedom; this *intrabeam scattering* can limit the luminosity lifetime of a hadron-hadron collider. Random noise in the radiofrequency acceleration system or in the magnet power system can lead to emittance dilution in the various degrees of freedom. Quantum fluctuations in the synchrotron radiation process excite transverse and longitudinal oscillations in electron rings. Another important source of emittance growth is errors in the transfer of a beam from one accelerator to another.

There is a distinct difference between electron accelerators and proton accelerators insofar as emittance preservation is concerned. As will be seen in Chapter 8, the radiation produced by an accelerated charge and the replenishment of the energy by the RF accelerating system causes the emittances of the beam in a synchrotron to vary with time. With appropriate choices of parameters, the system will damp oscillations in all three degrees of freedom. In electron synchrotrons, where radiation plays a dominant role, the emittances of the beam are virtually predetermined; while mechanisms for emittance growth are present, with quantized emission of energy being the dominant source, the damping of the oscillations will force the beam size to an equilibrium value. Proton synchrotrons are less forgiving, however. Since the radiation effects are many orders of magnitude smaller than in electron synchrotrons, other emittance growth mechanisms in proton synchrotrons can be of more serious concern for these devices. In this chapter we will hence concentrate on emittance preservation within hadron circular accelerators and beam transfers between such rings. The emittances of electron beams in circular accelerators will be discussed in Chapter 8.

We will discuss various emittance growth processes as applied to the transverse degrees of freedom. The longitudinal emittance of a bunched beam can also grow via many of the same processes; in fact, the longitudinal emittances of many modern proton synchrotrons are often increased intentionally, to provide Landau damping for instance. For cases where preservation of the longitudinal emittance is of interest, many of the same types of arguments as we will go through for transverse processes can also be applied. Some of these situations are presented in the problems at the end of the chapter.

Many of the sources of emittance dilution can be grouped into two important categories. The first contains mechanisms which cause single abrupt changes in the particle phase space distribution, resulting in a larger than desired phase space area demanded by the beam. The most common examples of such processes include steering and gradient errors encountered during the transfer of beams into a synchrotron. The second category contains mechanisms which "continuously" afflict the particles' oscillation amplitudes. In the following two sections we discuss two examples for each category. Other examples are left to the problems.

Though the effects mentioned thus far lead to emittance growth, methods have been developed to reduce beam emittances in all three degrees of freedom. Emittance reduction techniques were instrumental in the development of accumulator rings for antiprotons. The last section of this chapter is devoted to basic descriptions of stochastic beam cooling methods as they have been applied to hadron storage rings.

7.1 INJECTION MISMATCH

Modern high energy accelerator facilities employ a series of accelerators of various intermediate energies. In the design of beam transport systems between accelerators, the primary concern is to match the amplitude functions, dispersion functions, and of course the ideal beam trajectory coming from the first synchrotron to those of the second synchrotron. If a proper match is not provided, an increase in the transverse emittance will result.

As an example, consider a distribution of particles all of the ideal energy entering a synchrotron with the centroid of the distribution offset from the

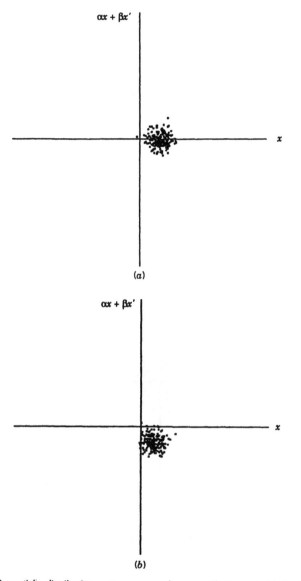

Figure 7.1. A particle distribution enters an accelerator with its centroid displaced from the ideal orbit. Due to nonlinearities in the transverse restoring forces, the betatron oscillation frequency depends upon oscillation amplitude. Over time, the motion decoheres and the distribution filaments. The result is an increased particle beam emittance.

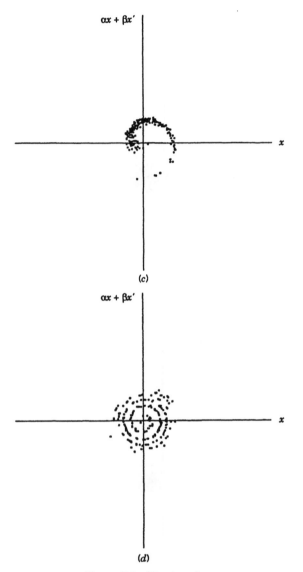

Figure 7.1. (*Continued*).

ideal orbit, as illustrated in Figure 7.1(a). If the synchrotron contained only ideal linear transverse restoring forces, the particles would undergo coherent betatron oscillations and the emittance of the beam itself would remain constant; however, the total phase space area which the beam explored would have effectively increased [Figure 7.1(b)]. In a more realistic accelerator, the magnetic fields will in general have nonlinear components, and thus

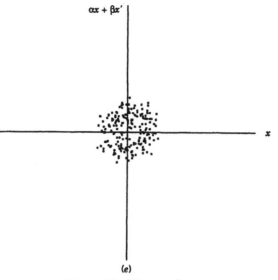

(e)

Figure 7.1. (*Continued*).

the oscillation frequency will depend upon the oscillation amplitude; the particle motion will eventually decohere and the beam distribution will filament, as shown in Figure 7.1(c) and (d). Finally, after enough time has elapsed, the phase space distribution might look like that in Figure 7.1(e), where cylindrical symmetry has been reestablished, but the emittance of the beam has increased.

For ease of computation we will assume in what follows that a particle's oscillation amplitude will remain invariant after injection into the synchrotron. Its tune may depend upon the amplitude, but whether the injected distribution continues to oscillate coherently or whether nonlinear fields cause the motion to decohere, the time average distribution of the particles will be the same provided the average is taken over a sufficiently long period.

Given a particle with an initial coordinate in phase space, the resulting time average distribution in the transverse coordinate may be obtained for that particle. Using this result, an expression for the final distribution of many particles, given their initial distribution, may be found. With this distribution function, the area in phase space which contains a certain fraction of the particles may be computed, as well as the rms beam size. Initial distributions generated by various forms of mismatch may then be inserted into these expressions to yield resulting time average distributions and emittance dilution factors.

We consider first the time average distribution of a single particle. If the transverse motion in one degree of freedom is observed at a particular longitudinal location s in the synchrotron, then the trajectory in x, p_x phase space, where $p_x \equiv \alpha x + \beta x'$, is

$$x^2 + p_x^2 = a^2, \tag{7.1}$$

where a is the amplitude of the particle motion at point s. The Courant-Snyder parameters α and β are evaluated at s.

Suppose a particle enters the accelerator and upon its first passage through point s the particle has phase space coordinates (x_0, p_{x0}). Upon subsequent revolutions about the machine, the particle will reappear at point s with phase space coordinates (x, p_x) which lie on a circle of radius $a = (x_0^2 + p_{x0}^2)^{1/2}$. The exact location on the circle after each revolution will depend upon the phase advance of the betatron oscillation for one complete revolution, which may in fact be dependent upon amplitude. Over a long period of time, the probability of finding the particle at a specific transverse displacement x may be computed. If x and p_x are parametrized by

$$x = a \cos \omega t, \tag{7.2}$$

$$p_x = -a \sin \omega t, \tag{7.3}$$

then the phase space distribution of the particle will be given by

$$g_1(x, p_x, t) \, dx \, dp_x = \delta(x - a \cos \omega t) \delta(p_x + a \sin \omega t) \, dx \, dp_x \tag{7.4}$$

where $\delta(u)$ is the Dirac δ-function. Integrating over p_x yields

$$g_2(x, t) \, dx = \delta(x - a \cos \omega t) \, dx. \tag{7.5}$$

To find the time average distribution in x, we may integrate $g_2(x, t)$ over a cycle of period $\tau = 2\pi/\omega$. In fact, due to the symmetry of the problem, integration over half a period is sufficient, which yields

$$n_a(x) \, dx = dx \, \frac{2}{\tau} \int_0^{\tau/2} \delta(x - a \cos \omega t) \, dt \tag{7.6}$$

$$= dx \, \frac{2}{\tau} \int_{-a}^a \delta(x - u) \frac{du}{a\omega \left[1 - (u/a)^2\right]^{1/2}}, \tag{7.7}$$

or

$$n_a(x) \, dx = \frac{1}{\pi a} \frac{dx}{\sqrt{1 - (x/a)^2}}. \tag{7.8}$$

Given the initial condition in transverse phase space (x_0, p_{x0}), over a long period of time the probability of finding the particle between x and $x + dx$ is $n_a(x) \, dx$.

Now given an initial distribution of particles $n_0(x, p_x) \, dx \, dp_x$ within the synchrotron at location s, then the resulting time average distribution of the particles may be found. Switching to polar coordinates, the number of particles which are located within a circle of radius a is given by

$$f(a) = \int_0^{2\pi} \int_0^a n_0(r, \theta) r \, dr \, d\theta, \tag{7.9}$$

and the number of particles between two circles of radii a and $a + da$ is

$$\frac{\partial f(a)}{\partial a} \, da = da \int_0^{2\pi} n_0(a, \theta) a \, d\theta. \tag{7.10}$$

Thus, the contribution of a particular ring of radius a and thickness da to the resulting time average distribution in x is

$$n_a(x) \, dx = \frac{1}{\pi} \frac{dx \, da}{\sqrt{1 - (x/a)^2}} \int_0^{2\pi} n_0(a, \theta) \, d\theta. \tag{7.11}$$

Upon adding up all contributions due to all pertinent rings (i.e., $a \geq |x|$), the resulting time average distribution in x will be

$$n(x) \, dx = \frac{dx}{\pi} \int_{|x|}^\infty \int_0^{2\pi} \frac{n_0(a, \theta)}{\sqrt{1 - (x/a)^2}} \, d\theta \, da. \tag{7.12}$$

Using this equation, the resulting time average distribution of particles in one degree of freedom may be computed given the initial distribution of particles delivered by the beamline. A perfect match of the beamline to the synchrotron would produce a resulting time average distribution of $n(x) \, dx = \int n_0(x, p_x) \, dp_x$.

The variance of the time average distribution can be obtained by integrating $\int x^2 n(x) \, dx$. An easier way of arriving at the average value of x^2 is to consider the symmetry of the distribution. The particle's trajectory in phase space is a circle, given by $x^2 + p_x^2 = a^2$. Averaging over time, we see that $\langle x^2 \rangle + \langle p_x^2 \rangle = a^2$, where the angle brackets denote time averages. But because the phase space trajectory is circular, $\langle x^2 \rangle = \langle p_x^2 \rangle$ and thus

$$\langle x^2 \rangle = a^2/2. \tag{7.13}$$

Hence, while the average position of the particle is $x = 0$, its rms position will be $\sigma \equiv x_{rms} = a/\sqrt{2}$.

With these general results for the time average distribution of a single particle we can now address questions about mismatches of beams of particles upon entering a synchrotron.

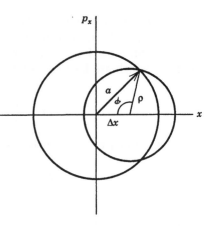

Figure 7.2. A particle distribution enters an accelerator with its centroid displaced by an amount Δx from the ideal orbit. A particle whose amplitude would have been ρ, had the central trajectory been matched, now oscillates with an amplitude a.

7.1.1 Steering Errors

Consider the effect of a steering error such as might occur at injection. The situation is illustrated in Figure 7.2. The origin of coordinates is on the design trajectory of the accelerator, but the beam enters with its centroid offset by an amount Δx in position and an amount $\Delta x'$ in slope. With respect to the centroid, the position of a particle may be characterized by the polar coordinates ρ, ϕ. However, upon entrance into the accelerator, the particle will undergo a betatron oscillation with amplitude a about the design orbit. The x, p_x axes may be rotated through an angle Θ given by

$$\tan \Theta = \frac{\Delta p_x}{\Delta x} = \frac{\beta \, \Delta x' + \alpha \, \Delta x}{\Delta x}, \tag{7.14}$$

so that the problem is equivalent to one in which the incoming distribution is displaced only in position by an amount

$$\Delta x_{eq} \equiv \sqrt{\Delta x^2 + (\beta \, \Delta x' + \alpha \, \Delta x)^2}. \tag{7.15}$$

From now on, Δx will be used to represent Δx_{eq}.

Let us assume that the incoming distribution is Gaussian with the form

$$n_0(x, p_x) \, dx \, dp_x = \frac{1}{2\pi\sigma_0^2} e^{-[(x - \Delta x)^2 + p_x^2]/2\sigma_0^2} \, dx \, dp_x \tag{7.16}$$

and compute the final time average distribution due to a position mismatch, $n_x(x) \, dx$. Switching to polar coordinates,

$$n_x(x) \, dx = \frac{dx}{2\pi^2 \sigma_0^2} \int_{|x|}^{+\infty} \int_0^{2\pi} \frac{e^{-[a^2 + \Delta x^2 - 2a \, \Delta x \cos \theta]/2\sigma_0^2}}{\sqrt{1 - (x/a)^2}} \, d\theta \, da, \tag{7.17}$$

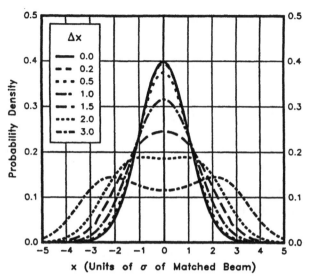

Figure 7.3. Particle distribution resulting from steering error at injection.

or, reducing the expression to a single integral,

$$n_x(x) \, dx = \frac{dx}{\pi\sigma_0^2} \int_{|x|}^{+\infty} e^{-(a-\Delta x)^2/2\sigma_0^2} \frac{e^{-|a\,\Delta x|/\sigma_0^2} I_0\!\left(a\,\Delta x/\sigma_0^2\right)}{\sqrt{1-(x/a)^2}} \, da, \quad (7.18)$$

where $I_0(z)$ is the modified Bessel function of order zero.[1] Numerical integration of this expression yields the curves shown in Figure 7.3. If the injected beam is displaced by more than about twice the standard deviation of the initial particle distribution, the resulting time average distribution exhibits a double hump.

The variance of the resulting particle distribution can be found using Equation 7.13. A single particle with initial coordinates corresponding to an amplitude a will provide a contribution of $a^2/2$ to the variance of the resulting time average distribution in x. From Figure 7.2 we see that

$$a^2 = \rho^2 + \Delta x^2 - 2\rho\,\Delta x \cos\phi. \qquad (7.19)$$

An initial distribution having a variance σ_0^2 and which has rotational symmetry about the point $(\Delta x, 0)$ can be written in the form

$$n_0(\rho, \phi) = f(\rho)/2\pi. \qquad (7.20)$$

---

[1]See, for example, M. Abramowitz, and I. A. Stegun, *Handbook of Mathematical Functions*, Dover, New York, 1970, p. 374.

The variance of the time average distribution can thus be computed as

$$\sigma^2 \equiv \langle x^2 \rangle = \int \frac{a^2}{2} n_0 \, d\Sigma \tag{7.21}$$

$$= \int \frac{\rho^2}{2} n_0 \, d\Sigma + \int \frac{\Delta x^2}{2} n_0 \, d\Sigma - \int \rho \Delta x \cos \phi n_0 \, d\Sigma$$

$$= \int \frac{\rho^2}{2} n_0 \, d\Sigma + \frac{\Delta x^2}{2} - \frac{\Delta x}{2\pi} \int \rho^2 f(\rho) \, d\rho \int \cos \phi \, d\phi \tag{7.22}$$

where $d\Sigma$ is a differential element of phase space area. While the third term is zero, the first term is just

$$\int \frac{\rho^2}{2} n_0 \, d\Sigma = \sigma_0^2, \tag{7.23}$$

which is the variance of the incoming distribution—that is, the variance the final distribution would have had if there were no mismatch. Thus, the variance of the resulting distribution is simply

$$\sigma^2 = \sigma_0^2 + \tfrac{1}{2} \Delta x^2. \tag{7.24}$$

While the distribution functions plotted in Figure 7.3 assume an initial Gaussian distribution, the expression for the variance in Equation 7.24 is completely general for any initial distribution with cylindrical symmetry in x, p_x phase space.

In Chapter 3 we derived an expression for the emittance of a beam distribution which is Gaussian in the transverse coordinate. If the incoming distribution were Gaussian, the resulting time average distribution due to an injection position mismatch would not be. However, provided that the mismatch is not too large, the area in phase space which will contain the beam is larger than the incoming emittance by the ratio

$$\frac{\epsilon}{\epsilon_0} = 1 + \frac{1}{2} \left(\frac{\Delta x}{\sigma_0} \right)^2. \tag{7.25}$$

Choosing one of the many definitions of the normalized emittance, such as $\epsilon_n = \pi \sigma^2 (\gamma v/c)/\beta$, we see that a steering error at injection will generate an increase in the beam emittance by an amount

$$\Delta \epsilon_n = \frac{\pi (\gamma v/c) \, \Delta \sigma^2}{\beta} = \frac{\pi (\gamma v/c)}{2} \frac{\Delta x^2 + (\beta \, \Delta x' + \alpha \, \Delta x)^2}{\beta} \tag{7.26}$$

where β and α are the Courant-Snyder parameters at the location of the observed trajectory errors and here the expression includes both errors in position and slope. It is interesting to note that this result is independent of the incoming beam size.

A mismatch of the dispersion function which is delivered to a synchrotron from a beamline can be handled analogously, since this is simply a mismatch of the trajectories of off-momentum particles. For a particle of momentum $p + \Delta p$, where p is the ideal momentum, the equilibrium orbit lies on the phase space point $(x, p_x) = (D \Delta p/p, (\beta D' + \alpha D) \Delta p/p)$, where D is the dispersion function. A mismatch of the incoming dispersion function to that of the accelerator will result in a steering error $\Delta x = \Delta D \Delta p/p$ and $\Delta x' = \Delta D' \Delta p/p$, which will result in an increased transverse emittance.

Without going through the same steps as before, we state the results for the time average distribution function due to a mismatch of the dispersion function. Let D be the dispersion function of the accelerator at the observation point, and ΔD_e be the deviation of that value delivered by the beamline:

$$\Delta D_e \equiv \left[\Delta D^2 + (\beta \Delta D' + \alpha \Delta D)^2 \right]^{1/2}. \tag{7.27}$$

Then

$$n_D(x) \, dx = \frac{dx}{\sqrt{2\pi^3} \, \sigma_0^2} \int_{-\infty}^{+\infty} \int_{|x - D\sigma_p\delta|}^{+\infty} e^{-\delta^2/2} e^{-(a - |\Delta D_e \, \sigma_p \delta|)^2/2\sigma_0^2}$$

$$\times \frac{e^{-|a \, \Delta D_e \, \sigma_p \delta| / \sigma_0^2} I_0 \left(a \, \Delta D_e \, \sigma_p \delta / \sigma_0^2 \right)}{\sqrt{1 - \left(\dfrac{x - D\sigma_p\delta}{a} \right)^2}} \, da \, d\delta. \tag{7.28}$$

Here $\delta \equiv (\Delta p/p)/\sigma_p$, where σ_p is the rms value of the relative momentum deviation; and we are assuming once again Gaussian distributions both in the initial transverse beam dimension and in momentum.

Note that for $\Delta D_e = 0$, the distribution function n_D becomes

$$n_D(x) \, dx = \frac{1}{\sqrt{2\pi (\sigma_0^2 + D^2\sigma_p^2)}} e^{-x^2/2(\sigma_0^2 + D^2\sigma_p^2)} \, dx, \tag{7.29}$$

which is a Gaussian distribution with variance $\sigma_0^2 + D^2\sigma_p^2$. From Equation 3.138 we know that the variance of the beam distribution at a location in a beamline or synchrotron where the dispersion function is nonzero is given by $\sigma^2 = \sigma_t^2 + D^2\sigma_p^2$, where σ_t contains the transverse emittance. The result of a dispersion function mismatch is to increase the transverse emittance and hence increase σ_t. To see this effect, we examine the resulting time average distribution with D set to zero; the total variance of the distribution which one would actually observe would be $D^2\sigma_p^2$ plus the variance of $n_D(D = 0)$.

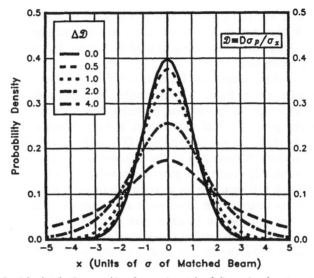

Figure 7.4. Particle distribution resulting from mismatch of dispersion function at injection. The curves are drawn for a point where the dispersion function of the synchrotron is zero.

Figure 7.4 shows the distribution $n_D(D = 0)$ for several values of the dispersion mismatch. The severity of the emittance dilution depends upon both ΔD_e and σ_p, as it must. If all of the particles were of the exact same momentum, the beam size would not increase no matter how large a value for ΔD_e was obtained. Likewise, any small deviation from the ideal dispersion function significantly affects the emittance of a beam which has a large enough momentum spread.

The variance of the distribution $n_D(D = 0)$ is given by

$$\sigma^2 = \sigma_0^2 + \tfrac{1}{2} \Delta D^2 \sigma_p^2, \tag{7.30}$$

and likewise the change in the normalized emittance, under the same assumptions of sufficiently small mismatch, is

$$\Delta\epsilon_n = \frac{\pi(\gamma v/c)}{2} \frac{\Delta D^2 + (\beta\,\Delta D' + \alpha\,\Delta D)^2}{\beta}\sigma_p^2. \tag{7.31}$$

7.1.2 Focusing Errors

The treatment for an amplitude function mismatch can be carried out in a similar fashion. To begin with, we must differentiate between the amplitude function which is being delivered by the beamline and the periodic amplitude

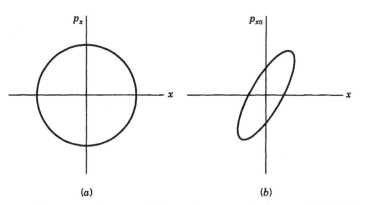

Figure 7.5. A phase space trajectory which is circular as viewed in terms of the beamline lattice functions (a) will in general appear elliptical when viewed in terms of the ring lattice functions (b).

function of the synchrotron. Suppose β and α are the Courant-Snyder parameters as delivered by the beamline to a particular point in an accelerator, and β_0, α_0 are the periodic lattice functions of the ring at that point. A particle with trajectory (x, x') can be viewed in the $(x, \beta x' + \alpha x) \equiv (x, p_x)$ phase space corresponding to the beamline functions, or in the $(x, \beta_0 x' + \alpha_0 x) \equiv (x, p_{x0})$ phase space corresponding to the lattice functions of the ring. If the phase space motion lies on a circle in the beamline view, then it will lie on an ellipse in the ring view, as indicated in Figure 7.5.

The equation of the ellipse can be obtained by noting that

$$x' = \frac{p_x - \alpha x}{\beta} = \frac{p_{x0} - \alpha_0 x}{\beta_0}, \tag{7.32}$$

or

$$p_x = \frac{\beta}{\beta_0} p_{x0} + \left(\alpha - \frac{\beta}{\beta_0} \alpha_0 \right) x \tag{7.33}$$

$$\equiv \beta_r p_{x0} + \Delta\alpha_r x. \tag{7.34}$$

If the equation of the circle in the beamline view is

$$x^2 + p_x^2 = \beta A^2, \tag{7.35}$$

where A^2 is the Courant-Snyder invariant, then the equation of the ellipse in the ring system will be

$$(1 + \Delta\alpha_r^2) x^2 + 2\beta_r \Delta\alpha_r x p_{x0} + \beta_r^2 p_{x0}^2 = \beta A^2. \tag{7.36}$$

It will be useful to rotate the coordinate axes so that they correspond to the major and minor axes of the ellipse. This amounts to rotating through an angle Θ so that the cross term in the equation of the ellipse is eliminated. The angle is given by

$$\tan 2\Theta = \frac{2\beta_r \Delta\alpha_r}{1 + \Delta\alpha_r^2 - \beta_r^2}, \tag{7.37}$$

and the resulting equation in the rotated coordinates $x_e \equiv x \cos \Theta + p_x \sin \Theta$, $p_{xe} \equiv -x \sin \Theta + p_x \cos \Theta$ will be

$$\frac{1}{2}\left[(1 + \Delta\alpha_r^2 + \beta_r^2) + \sqrt{(1 + \Delta\alpha_r^2 - \beta_r^2)^2 + 4(\beta_r \Delta\alpha_r)^2}\,\right]x_e^2$$

$$+ \frac{1}{2}\left[(1 + \Delta\alpha_r^2 + \beta_r^2) - \sqrt{(1 + \Delta\alpha_r^2 - \beta_r^2)^2 + 4(\beta_r \Delta\alpha_r)^2}\,\right]p_{xe}^2$$

$$= \beta A^2. \tag{7.38}$$

But the expression under the radical can be simplified:

$$\left(1 + \Delta\alpha_r^2 - \beta_r^2\right)^2 + 4(\beta_r \Delta\alpha_r)^2 = \left(1 + \Delta\alpha_r^2 + \beta_r^2\right)^2 - 4\beta_r^2. \tag{7.39}$$

We also note that

$$1 + \Delta\alpha_r^2 + \beta_r^2 = 1 + (\alpha - \beta_r\alpha_0)^2 + \beta_r^2$$

$$= \beta_r\left[\frac{1 + \alpha^2}{\beta_r} + \beta_r(1 + \alpha_0^2) - 2\alpha\alpha_0\right]$$

$$= \beta_r\left[\beta_0\frac{1 + \alpha^2}{\beta} + \beta\frac{1 + \alpha_0^2}{\beta_0} - 2\alpha\alpha_0\right]$$

$$= \beta_r[\beta_0\gamma + \beta\gamma_0 - 2\alpha\alpha_0]$$

$$= 2\beta_r F, \tag{7.40}$$

where

$$F \equiv \tfrac{1}{2}(\beta_0\gamma + \beta\gamma_0 - 2\alpha\alpha_0). \tag{7.41}$$

Using Equations 7.39 and 7.40, Equation 7.38 becomes

$$\frac{F + \sqrt{F^2 - 1}}{\beta_0 A^2}x_e^2 + \frac{F - \sqrt{F^2 - 1}}{\beta_0 A^2}p_{xe}^2 = 1, \tag{7.42}$$

which is the standard form of the equation of an ellipse in terms of its major and minor axes.

Finally, we recognize that

$$\frac{1}{F - \sqrt{F^2 - 1}} = F + \sqrt{F^2 - 1} \qquad (7.43)$$

and so the equation of the ellipse is of the form

$$b_r x_e^2 + \frac{1}{b_r} p_{xe}^2 = \beta_0 A^2, \qquad (7.44)$$

where

$$b_r \equiv F + \sqrt{F^2 - 1}. \qquad (7.45)$$

Note that for the special case where $\alpha = \alpha_0 = 0$ and $\beta \neq \beta_0$, then $\Theta = 0$, $F = [(\beta/\beta_0) + (\beta_0/\beta)]/2$, and $b_r = \beta/\beta_0 = (\beta_0 + \Delta\beta)/\beta_0 = 1 + \Delta\beta/\beta_0$. That is, $b_r - 1 = \Delta\beta/\beta_0$ represents the "amplitude" of a "beta function" mismatch.

We can now proceed to look at the time average distribution which results from an initial distribution created by an amplitude function mismatch. If the incoming distribution is a Gaussian with cylindrical symmetry when viewed in terms of the beamline lattice functions, then in the phase space corresponding to the ring lattice functions, the distribution will have the form

$$n_0(x, p_x) \, dx \, dp_x = \left(\frac{e^{-x^2/2b_r\sigma_0^2}}{\sqrt{2\pi b_r \sigma_0^2}} \right) \left(\frac{e^{-b_r p_x^2/2\sigma_0^2}}{\sqrt{2\pi\sigma_0^2/b_r}} \right) \qquad (7.46)$$

where the subscripts e have been suppressed to simplify notation. Here, σ_0 corresponds to the rms displacement the particle distribution would have if the lattice functions were perfectly matched.

Upon transforming to polar coordinates, the integral for the resulting time average distribution due to an amplitude function mismatch becomes

$$n_\beta(x) \, dx = \frac{dx}{2\pi^2\sigma_0^2} \int_{|x|}^{+\infty} \int_0^{2\pi} \frac{\exp\left[-\frac{a^2}{2b_r\sigma_0^2} (\cos^2\theta + b_r^2 \sin^2\theta) \right]}{\sqrt{1 - (x/a)^2}} \, d\theta \, da, \qquad (7.47)$$

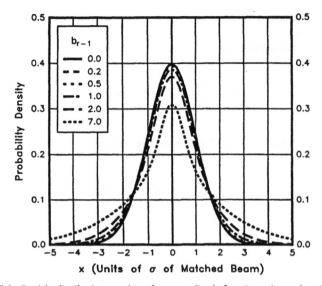

Figure 7.6. Particle distribution resulting from amplitude function mismatch at injection.

or

$$n_\beta(x)\,dx = \frac{dx}{\pi\sigma_0^2}\int_{|x|}^{+\infty} e^{-a^2/2b_r\sigma_0^2}\,\frac{\exp\left[-\left(\dfrac{a^2}{4}\dfrac{b_r^2-1}{b_r}\right)\right]I_0\left(\dfrac{a^2}{4}\dfrac{b_r^2-1}{b_r}\right)}{\sqrt{1-(x/a)^2}}\,da.$$

(7.48)

Figure 7.6 shows the resulting distribution. In contrast to the distributions resulting from steering errors, the centroid of the beam is hardly disturbed, and hence, as can be seen, the amplitude function must be greatly mismatched to produce a significant increase in the variance of the distribution.

Once again the variance of the resulting particle distribution can be found using Equation 7.13. If the initial coordinates of a particular particle are given by (x_0, p_{x0}) where

$$b_r x_0^2 + \frac{1}{b_r}p_{x0}^2 = \beta_0 A^2,$$

(7.49)

then this particle will commence describing a circular trajectory in phase space on subsequent passages through the synchrotron, where the radius of the phase space trajectory is given by $a = \sqrt{x_0^2 + p_{x0}^2}$. Upon averaging over the entire distribution, we have

$$\langle x_0^2 \rangle = b_r \sigma_0^2, \tag{7.50}$$

$$\langle p_{x0}^2 \rangle = \frac{1}{b_r}\sigma_0^2, \tag{7.51}$$

and hence

$$\langle a^2 \rangle = \langle x_0^2 \rangle + \langle p_{x0}^2 \rangle \tag{7.52}$$

$$= b_r \sigma_0^2 + \frac{1}{b_r}\sigma_0^2. \tag{7.53}$$

It follows that the resulting distribution will have variance

$$\sigma^2 = \frac{\langle a^2 \rangle}{2} = \frac{b_r^2 + 1}{2 b_r}\sigma_0^2. \tag{7.54}$$

But $b_r = F + \sqrt{F^2 - 1}$ and hence $b_r^2 + 1 = 2Fb_r$. Therefore, upon averaging over time, the variance of the distribution will be increased by a factor

$$\sigma^2/\sigma_0^2 = F = \tfrac{1}{2}(\beta\gamma_0 + \beta_0\gamma - 2\alpha\alpha_0). \tag{7.55}$$

This expression can be made to look more like the expressions obtained for steering errors by rewriting it as (see the Problems)

$$F = 1 + \tfrac{1}{2}|\det \Delta J| \tag{7.56}$$

where J is the 2×2 matrix containing the Courant-Snyder parameters:

$$J = \begin{pmatrix} \alpha & \beta \\ -\gamma & -\alpha \end{pmatrix}. \tag{7.57}$$

For the case where the slope of the amplitude function is matched and equal to zero, we have

$$\frac{\sigma^2}{\sigma_0^2} = 1 + \frac{1}{2}\left(\frac{\Delta\beta/\beta_0}{\sqrt{1 + \Delta\beta/\beta_0}} \right)^2. \tag{7.58}$$

It is interesting to note that the change in emittance (or σ^2) generated by an amplitude function mismatch is proportional to the incoming emittance, in contrast to the effect of a steering error, where the emittance increase is independent of the initial emittance.

Table 7.1 summarizes the emittance dilution factors to injection amplitude function, dispersion function, and steering errors.

Table 7.1. Transverse emittance dilution factors;
see text for explanations.

Amplitude function mismatch:

$$\frac{\sigma^2}{\sigma_0^2} = 1 + \tfrac{1}{2}|\det \Delta J|$$

Dispersion function mismatch:

$$\frac{\sigma^2}{\sigma_0^2} = 1 + \frac{1}{2}\left(\frac{\Delta D^2 + (\beta\,\Delta D' + \alpha\,\Delta D)^2}{\sigma_0^2}\right)\sigma_p^2$$

Injection steering error:

$$\frac{\sigma^2}{\sigma_0^2} = 1 + \frac{1}{2}\left(\frac{\Delta x^2 + (\beta\,\Delta x' + \alpha\,\Delta x)^2}{\sigma_0^2}\right)$$

7.2 DIFFUSION PROCESSES

In this section we wish to discuss the general behavior of particle distributions in the presence of mechanisms which continuously stimulate emittance growth. Let us suppose we have an initial distribution of particles in transverse phase space, $f_0(x, x')$, where x is the transverse coordinate and x' is the slope of a particle's trajectory, $x' = dx/ds$, as observed at some particular point in the accelerator. In the absence of such mechanisms, we assume that this distribution will not change with time, and so

$$f(x, x', t) = f_0(x,x'). \tag{7.59}$$

However, if some process is randomly altering the betatron amplitudes of the particles in the beam, the extent of the distribution will grow with time and $f(x, x', t)$ will satisfy the diffusion equation.

To understand the diffusion equation, imagine a system of particles, constrained to move with one degree of freedom, with density function $f(x)$. Let J be the average number of particles per unit area crossing a plane perpendicular to the x-direction per unit time. We now look at a region around x, as sketched in Figure 7.7, and consider the number of particles flowing into and out of this region.

The number flowing into the region bounded by x and $x + \Delta x$ and with cross-sectional area A in the small time interval Δt is $AJ(x)\,\Delta t$. The

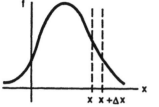

Figure 7.7. Diffusion of particles across a boundary at x.

number flowing out is given by $AJ(x + \Delta x)\,\Delta t$. Thus,

$$\frac{\partial}{\partial t}(fA\,\Delta x) = AJ(x) - AJ(x + \Delta x), \qquad (7.60)$$

or

$$\frac{\partial f}{\partial t} = -\frac{\partial J}{\partial x}. \qquad (7.61)$$

Now if f is uniform, the migration of particles due to some random process into the region between x and $x + \Delta x$ will equal the average flow of particles out of this region, and so $J(x)$ would be zero. If the density function had a greater value at x than at $x + \Delta x$ (i.e., if f has a nonzero gradient), then more particles are apt to wander into the region from the left than from the right. That is, one would expect J to be proportional (to good approximation) to the rate of change of f with respect to x:

$$J = -C\frac{\partial f}{\partial x}, \qquad (7.62)$$

where C is a constant of proportionality. Substituting Equation 7.62 into Equation 7.61, we obtain the diffusion equation in one degree of freedom:

$$\frac{\partial f}{\partial t} = C\frac{\partial^2 f}{\partial x^2}. \qquad (7.63)$$

The more general three-dimensional form of the diffusion equation is

$$\frac{\partial f}{\partial t} = C\,\nabla^2 f. \qquad (7.64)$$

For the case of diffusion in one of the transverse degrees of freedom of a particle distribution in an accelerator, it is again useful to consider the distribution in x, p_x phase space. We consider only initial distributions with cylindrical symmetry in this phase space. Let r be the amplitude of a particle's transverse oscillation: $r^2 = x^2 + p_x^2 = x^2 + (\beta x' + \alpha x)^2$. Then the density function, in polar coordinates, satisfies

$$\frac{\partial f}{\partial t} = C \, \nabla^2 f = C \frac{1}{r} \frac{\partial}{\partial r}\left(r \frac{\partial f}{\partial r}\right). \tag{7.65}$$

We can see the significance of the diffusion constant C by multiplying Equation 7.65 by r^2 and integrating over all of phase space. We find

$$\int r^2 \frac{\partial f}{\partial t} \, r \, dr = C \int r \frac{\partial}{\partial r}\left(r \frac{\partial f}{\partial r}\right) r \, dr, \tag{7.66}$$

$$\frac{\partial}{\partial t} \int r^2 fr \, dr = C \int r^2 \, d\left(r \frac{\partial f}{\partial r}\right),$$

$$\frac{\partial}{\partial t}\langle r^2 \rangle = C \left[\left(r^3 \frac{df}{dr}\right)_0^\infty - \int \left(r \frac{df}{dr}\right)(2r \, dr)\right]$$

$$= -2C \int r^2 \, df$$

$$= -2C \left[(r^2 f)\big|_0^\infty - \int f \cdot 2r \, dr\right]$$

$$= 4C \int fr \, dr$$

$$= 4C, \tag{7.67}$$

or

$$C = \frac{1}{4} \frac{\partial}{\partial t}\langle r^2 \rangle. \tag{7.68}$$

We thus see that C is related to the time rate of change of the emittance of the beam. We therefore transform coordinates to involve the Courant-Snyder invariant $W \equiv [x^2 + (\beta x' + \alpha x)^2]/\beta = r^2/\beta$. So we may write $f = f(W, t)$ and note that the variable W is independent of longitudinal location within the accelerator. The diffusion equation then becomes

$$\frac{\partial f}{\partial t} = \frac{4C}{\beta} \frac{\partial}{\partial W}\left(W \frac{\partial f}{\partial W}\right). \tag{7.69}$$

To allow for processes which may occur at various points along the circumference, we can define a new diffusion constant

$$R = \frac{d}{dt}\left\langle \frac{r^2}{\beta}\right\rangle = \frac{d}{dt}\langle W \rangle \tag{7.70}$$

and write our diffusion equation as

$$\frac{\partial f}{\partial t} = R \frac{\partial}{\partial W}\left(W \frac{\partial f}{\partial W}\right). \tag{7.71}$$

To proceed, we define two quantities:

$$Z = \frac{W}{W_a}, \tag{7.72}$$

$$\tau = \left(\frac{R}{W_a}\right)t, \tag{7.73}$$

where W_a is the Courant-Snyder invariant corresponding to the limiting aperture of the accelerator (i.e., the admittance). If a is the half aperture at a location where the amplitude function has the value β, then $W_a = a^2/\beta$. Notice that Z and τ are both dimensionless quantities.

In terms of Z and τ, the problem reduces to

$$\frac{\partial f}{\partial \tau} = \frac{\partial}{\partial Z}\left(Z \frac{\partial f}{\partial Z}\right) \tag{7.74}$$

subject to the boundary conditions

$$f(Z,0) = f_0(Z), \tag{7.75}$$

$$f(1,\tau) = 0. \tag{7.76}$$

The solution of the above differential equation is

$$f(Z,\tau) = \sum_n c_n J_0(\lambda_n \sqrt{Z})e^{-\lambda_n^2 \tau/4} \tag{7.77}$$

with

$$c_n = \frac{1}{J_1(\lambda_n)^2}\int_0^1 f_0(Z)J_0(\lambda_n \sqrt{Z})\,dZ, \tag{7.78}$$

where λ_n is the nth zero of the Bessel function $J_0(z)$ ($\lambda_1 = 2.405$, $\lambda_2 = 5.520,\ldots$).

We now consider a particular form of the initial distribution, namely a bi-Gaussian in x, x' phase space. For this situation, the function f_0 will be

$$f_0(x, p_x) \, dx \, dp_x = f_0(r) r \, dr \tag{7.79}$$

$$= \frac{1}{\sigma^2} e^{-r^2/2\sigma^2} r \, dr \tag{7.80}$$

$$= \tfrac{1}{2} e^{-r^2/2\sigma^2} \, d(r^2/\sigma^2), \tag{7.81}$$

or

$$f_0(Z) \, dZ = \frac{a^2}{2\sigma^2} e^{-(a^2/2\sigma^2)Z} \, dZ. \tag{7.82}$$

So the coefficients c_n become

$$c_n = \frac{\alpha}{J_1(\lambda_n)^2} \int_0^1 e^{-\alpha Z} J_0(\lambda_n \sqrt{Z}) \, dZ, \tag{7.83}$$

where $\alpha = a^2/2\sigma^2$. If the entire initial beam distribution lies well within the aperture so that the integrand is sufficiently near zero before Z approaches 1, i.e., if α is greater than about 5, then the c_n's may be approximated by

$$c_n = \frac{1}{J_1(\lambda_n)^2} e^{-z_n} (\cosh z_n - \sinh z_n) \tag{7.84}$$

$$= \frac{1}{J_1(\lambda_n)^2} e^{-2z_n}, \tag{7.85}$$

where

$$z_n = \frac{\lambda_n^2}{4} \left(\frac{\sigma}{a} \right)^2. \tag{7.86}$$

If the initial distribution does not satisfy the above condition, then the c_n integrals may be performed numerically.

The development of the particle distribution with time, as well as the total beam intensity as a function of time, may now be computed. By integrating

$f(Z, \tau)$ over the range of Z, the number of particles $N(\tau)$ may be obtained, namely,

$$N(\tau) = \int_0^1 f(Z, \tau) \, dZ \tag{7.87}$$

$$= \int_0^1 \sum_n c_n J_0(\lambda_n \sqrt{Z}) e^{-\lambda_n^2 \tau/4} \, dZ, \tag{7.88}$$

$$= \sum_n c_n \int_0^1 J_0(\lambda_n \sqrt{Z}) e^{-\lambda_n^2 \tau/4} \, dZ \tag{7.89}$$

or

$$N(\tau) = 2 \sum_n \frac{c_n}{\lambda_n} J_1(\lambda_n) e^{-\lambda_n^2 \tau/4}. \tag{7.90}$$

For $\sigma \ll a$, this becomes

$$N(\tau) = 2 \sum_n \frac{1}{\lambda_n J_1(\lambda_n)} \exp\left\{ -\frac{\lambda_n^2}{4} \left[\tau + 2\left(\frac{\sigma}{a}\right)^2 \right] \right\}. \tag{7.91}$$

Figure 7.8(a) shows how $f(Z, \tau)$ varies with time for the case $\sigma/a = 0.20$. The particle distribution grows in transverse size until it reaches the aperture ($Z = 1$), at which time the area under the curve quickly begins to decrease. The intensity $N(\tau)$ for this same case is displayed in Figure 7.8(b). From $\tau = 0$ to $\tau \approx 0.1$ the intensity is nearly constant. Upon reaching the aperture limit, the intensity rapidly falls off until $\tau \approx 0.5$, where the final lifetime $4/\lambda_1^2$ is reached.

When a substantial fraction of the initial Gaussian distribution lies outside the aperture limit (all particles with $Z > 1$ being lost immediately), the rather flat region of $N(Z)$ for small Z disappears. Figure 7.9(a) shows the intensity vs. time for the case $\sigma/a = 0.2$. The lifetime, determined by

$$\tau_L(\tau) = -\frac{N(\tau)}{dN/d\tau}, \tag{7.92}$$

is shown in Figure 7.9(b) for the same two cases. The one case begins with a very long lifetime which then decreases to the value $4/\lambda_1^2$, while the other case begins with a very short lifetime which rapidly approaches its asymptotic value.

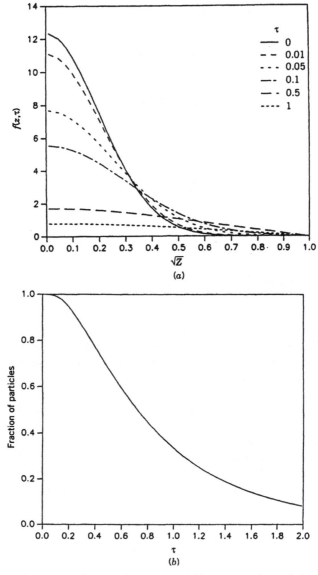

Figure 7.8. (a) Variation of density function f and (b) variation of particle beam intensity with time for an aperture at 5σ.

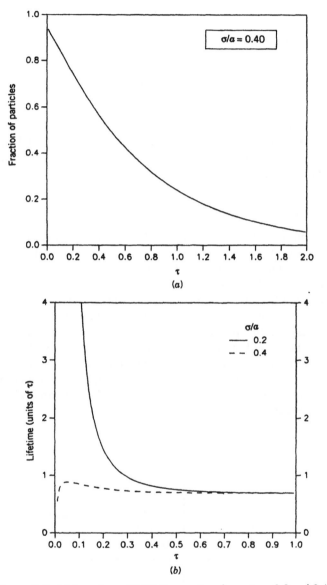

Figure 7.9. (a) Intensity and (b) lifetime vs. time for $\sigma / a = 0.2$ and 0.4.

The dimensionless quantity τ is related to time t by

$$\tau = \left(\frac{R}{W_a}\right)t,\qquad(7.93)$$

and hence the asymptotic lifetime is given by

$$t_L = \frac{4}{\lambda_1^2}\frac{W_a}{R}.\qquad(7.94)$$

If the emittance growth is caused by the changing of particle direction due to fluctuations in magnetic fields, elastic scattering off residual gas particles, etc., then the next step is to evaluate the diffusion constant R under these various circumstances.

7.2.1 RF Noise and Excitation of Oscillations

In this section, we wish to look at the development of betatron oscillations, and hence growth in emittance, if there is a sequence of random energy changes such as might be produced by noise in a radiofrequency accelerating system. We assume for this argument that each energy increment upon each passage through the accelerating station is uncorrelated with all other such occurrences. Now suppose the momentum of a particle changes abruptly by an amount Δp. If the particle were not already undergoing a synchrotron oscillation, one would now begin with initial conditions $\Delta E = \beta pcu$ and $\Delta\phi = 0$, where $u \equiv \Delta p/p$. If the dispersion D or its derivative D' is different from zero, a betatron oscillation will start with respect to the new off-momentum orbit with initial conditions $x = -Du$ and $x' = -D'u$, as indicated in Figure 7.10.

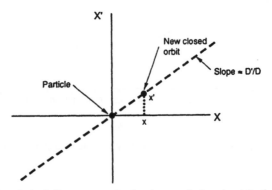

Figure 7.10. The dashed line represents the locus of closed orbits for various particle momenta.

Let's consider a sequence of fractional momentum changes $\{u_i\}$. In our familiar vector notation, after the first kick,

$$\begin{pmatrix} x_1 \\ x_1' \end{pmatrix} = -u_1 \begin{pmatrix} D \\ D' \end{pmatrix}. \tag{7.95}$$

After transformation by the single-turn matrix M, a second kick is delivered, so

$$\begin{pmatrix} x_2 \\ x_2' \end{pmatrix} = M \begin{pmatrix} x_1 \\ x_1' \end{pmatrix} - u_2 \begin{pmatrix} D \\ D' \end{pmatrix} \tag{7.96}$$

$$= -(u_1 M + u_2) \begin{pmatrix} D \\ D' \end{pmatrix}, \tag{7.97}$$

and after n turns

$$\begin{pmatrix} x_n \\ x_n' \end{pmatrix} = -(u_1 M^{n-1} + u_2 M^{n-2} + \cdots + u_{n-1} M + u_n) \begin{pmatrix} D \\ D' \end{pmatrix} \tag{7.98}$$

$$= -\sum_{m=1}^{n} u_m M^{n-m} \begin{pmatrix} D \\ D' \end{pmatrix}. \tag{7.99}$$

We're interested in emittance growth, so we should look at the Courant-Snyder invariant, which is given by

$$W = \frac{r^2}{\beta} = \gamma x^2 + 2\alpha xx' + \beta x'^2 = -\vec{x}^T S J \vec{x}, \tag{7.100}$$

where

$$J \equiv \begin{pmatrix} \alpha & \beta \\ -\gamma & -\alpha \end{pmatrix}, \tag{7.101}$$

$$S \equiv \begin{pmatrix} 0 & 1 \\ -1 & 0 \end{pmatrix}, \tag{7.102}$$

and A^T refers to the transpose of A.

Writing M in the form

$$M = I \cos \mu + J \sin \mu, \tag{7.103}$$

where μ is the phase advance for one revolution, each term in $W(n)$ will be of the form

$$-u_m u_k \vec{D}^T (I \cos \mu_1 + J^T \sin \mu_1) S J (I \cos \mu_2 + J \sin \mu_2) \vec{D}, \tag{7.104}$$

where $\mu_1 = (n-m)\mu$ and $\mu_2 = (n-k)\mu$. Noting that $J^T S = -SJ$ and $J^2 = -I$, this expression reduces to

$$\mathscr{H} u_m u_k \cos(m-k)\mu \tag{7.105}$$

with $\mathcal{H}$ defined as

$$\mathcal{H} = \gamma D^2 + 2\alpha DD' + \beta D'^2. \tag{7.106}$$

Then we have

$$W = \mathcal{H} \sum_{m,k}^{n} u_m u_k \cos(m - k)\mu. \tag{7.107}$$

Since we do not know the particular sequence of errors $\{u_i\}$, the best we can do is to average over a large ensemble of such sequences. Then

$$\langle W \rangle = \mathcal{H} \sum_{m,k}^{n} \cos(m - k)\mu \cdot \frac{1}{P} \sum_{p=1}^{P} u_{m,p} u_{k,p} \tag{7.108}$$

where P is the total number of sets in the ensemble. For uncorrelated kicks such as we would have from a truly random noise source,

$$\frac{1}{P} \sum_{p=1}^{P} u_{m,p} u_{k,p} = \delta_{mk} u_m^2 \tag{7.109}$$

and so

$$\langle W \rangle = \mathcal{H} \sum_{m=1}^{n} u_m^2 = n\mathcal{H}\langle u^2 \rangle, \tag{7.110}$$

which gives for the diffusion constant

$$R = \frac{d}{dt}\langle W \rangle = f_0 \mathcal{H}\langle u^2 \rangle = f_0 \mathcal{H} \frac{e^2 \langle v^2 \rangle}{\left(\dfrac{v}{c}\right)^4 E^2}, \tag{7.111}$$

where $\sqrt{\langle v^2 \rangle}$ is the rms voltage due to noise in the accelerating system and f_0 is the revolution frequency. In terms of a normalized emittance $\epsilon_N \equiv \pi\gamma(v/c)\sigma^2/\beta$,

$$\frac{d\epsilon_N}{dt} = \frac{\pi\gamma}{2} f_0 \mathcal{H} \frac{e^2 \langle v^2 \rangle}{\left(\dfrac{v}{c}\right)^3 E^2}, \tag{7.112}$$

where we note that $\sigma^2 = \langle x^2 \rangle = \langle r^2 \rangle/2$. Note that if the dispersion function and its derivative are both zero at the source of the noise, emittance growth can be avoided.

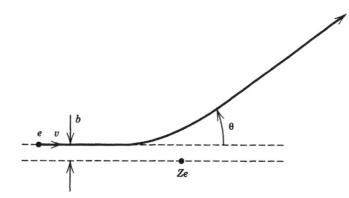

Figure 7.11. Coulomb scattering of electron from target nucleus.

7.2.2 Beam-Gas Scattering

As another example of a diffusion process we consider the small angle scattering of particles in the beam off of residual gas molecules in the vacuum chamber. We first consider the deflection of a single particle from a target nucleus and then the effect of multiple scattering off many such targets. From this we can obtain an expression for the time rate of change of the rms particle amplitude due to beam-gas scattering, and hence an emittance growth rate.

Coulomb Scattering Consider an incident particle of charge e approaching a target nucleus of charge Ze as shown in Figure 7.11. The particle approaches with a speed v and impact parameter b. The Coulomb repulsion (or attraction) changes the direction of motion of the particle by the angle θ. If the scattering angle is small, the transverse momentum $p_\perp$ that the particle acquires is given by

$$p_\perp = \int F_\perp \, dt = e \int E_\perp \, dt = \frac{e}{v} \int E_\perp \, dz \qquad (7.113)$$

$$= \frac{e}{v} \frac{1}{2\pi b} \int E_\perp \, dA = \frac{e}{v} \frac{1}{2\pi b} \frac{Ze}{\epsilon_0}, \qquad (7.114)$$

and

$$\theta = \frac{p_\perp}{p} = \frac{Ze^2}{2\pi \epsilon_0 p v b}. \qquad (7.115)$$

The differential cross section $d\sigma/d\Omega$ is the area $d\sigma$ presented by a target particle for scattering an incident particle into the solid angle $d\Omega$. For a particle approaching the target with impact parameter b, $d\sigma = 2\pi b\,db$. For small angles, $d\Omega = 2\pi\theta\,d\theta$. Hence, the differential cross section is

$$\frac{d\sigma}{d\Omega} = \left| \frac{2\pi b\,db}{2\pi\theta\,d\theta} \right|. \tag{7.116}$$

The absolute value is taken because both the area and the solid angle are positive quantities.

Using

$$b = \frac{Ze^2}{2\pi\epsilon_0 pv\theta}, \tag{7.117}$$

we have

$$\frac{db}{d\theta} = -\frac{Ze^2}{2\pi\epsilon_0 pv\theta^2}, \tag{7.118}$$

from which

$$\frac{d\sigma}{d\Omega} = \frac{b}{\theta}\frac{db}{d\theta} = 4\left(\frac{Ze^2}{4\pi\epsilon_0 pv} \right)^2 \frac{1}{\theta^4}. \tag{7.119}$$

This is the small angle limit of the famous Rutherford scattering cross section formula.

Multiple Coulomb Scattering   We next imagine a thin layer of material through which a particle passes, interacting with many atoms along its way. Upon each interaction, the particle's transverse coordinate is changed very little, but its direction of motion is altered according to the results of the section above. For the interaction with one scattering center, the variance of the particle's scattering angle is given by

$$\langle\theta^2\rangle_1 = \frac{\int \theta^2 (d\sigma/d\Omega)\,d\Omega}{\int (d\sigma/d\Omega)\,d\Omega}. \tag{7.120}$$

The limits of integration are

$$\theta_{min} \approx \frac{Ze^2}{2\pi\epsilon_0 pva}, \tag{7.121}$$

$$\theta_{max} \approx \frac{Ze^2}{2\pi\epsilon_0 pvR}, \tag{7.122}$$

where a is the "radius" of the target atom, and R is the "radius" of the target nucleus.

Using the Rutherford cross section above, we find that

$$\langle \theta^2 \rangle_1 = 2\theta_{min}^2 \ln(\theta_{max}/\theta_{min})$$

$$= 8Z^2 r_e^2 \left(\frac{m_e c^2}{pv} \right)^2 \frac{\ln(a/R)}{a^2} \tag{7.123}$$

where $r_e = e^2/(4\pi\epsilon_0 m_e c^2)$ is the classical radius of the electron. By adding up the contributions due to the scattering off the scattering centers within radius a of the particle's trajectory and through a thickness l of a material of density ρ and atomic weight A, we get

$$\langle \theta^2 \rangle = \frac{N_A}{A} \rho (l\pi a^2) \langle \theta^2 \rangle_1 \tag{7.124}$$

for the variance of the scattering angle distribution. Here N_A is Avagradro's number.

When dealing with radiation processes one often expresses lengths in units of the radiation length L_{rad}, given by

$$\frac{1}{L_{rad}} \equiv 2\alpha \frac{N_A}{A} \rho Z^2 r_e^2 \ln \frac{a}{R}, \tag{7.125}$$

where $\alpha \approx 1/137$ is the fine structure constant. In tables, the material density is often suppressed and "lengths" are expressed in units of grams per square centimeter.

The variance of the scattering angle distribution may now be written as

$$\langle \theta^2 \rangle = \frac{4\pi}{\alpha} \left(\frac{m_e c^2}{pv} \right)^2 \frac{l}{L_{rad}}$$

$$= \left(\frac{E_s}{pv} \right)^2 \frac{l}{L_{rad}}, \tag{7.126}$$

where $E_s = m_e c^2 \sqrt{4\pi/\alpha} = 21$ MeV. If we now consider the projection onto one particular transverse plane, then

$$\langle \theta^2 \rangle = \langle \theta_x^2 \rangle + \langle \theta_y^2 \rangle = 2\langle \theta_x^2 \rangle. \tag{7.127}$$

Therefore, the variance of the scattering angle in one transverse degree of freedom is given by

$$\langle \theta_x^2 \rangle = \left(\frac{15 \text{ MeV}}{pv} \right)^2 \frac{l}{L_{rad}}. \tag{7.128}$$

The Diffusion Constant   Finally, we use the result of the previous section to discuss the scattering of the beam particles off of the residual particles in the vacuum chamber. The average rate of change of the transverse scattering angle in one degree of freedom is given by

$$\frac{d}{dt}\langle\theta^2\rangle = \left(\frac{15\ \text{MeV}}{pv}\right)^2 \frac{c}{L_\text{rad}} \tag{7.129}$$

$$= \left(\frac{15\ \text{MeV}}{\gamma mv^2}\right)^2 \frac{c}{L_\text{rad}} \tag{7.130}$$

$$= (3.3 \times 10^{-7}/\text{sec})\frac{P[\mu\text{Torr}]}{\gamma^2}, \tag{7.131}$$

where the last expression is given for a proton beam, assuming air for the residual gas in the vacuum chamber, and assuming $v \approx c$. Here, $P[\mu\text{Torr}]$ is the average vacuum chamber pressure expressed in microtorrs.

The diffusion constant is $R = d\langle W\rangle/dt$ and $\Delta W = \Delta\{(x^2 + [\beta x' + \alpha x]^2)/\beta\} = (2\alpha x\, \Delta x' + \beta^2\, \Delta x'^2)/\beta$. So, when averaging over many scatterings,

$$R = \left\langle\beta\frac{d}{dt}\Delta x'^2\right\rangle = \langle\beta\rangle\langle\dot\theta^2\rangle, \tag{7.132}$$

and the emittance growth rate, using $\epsilon_N \equiv \pi\gamma(v/c)\sigma^2/\beta$, is

$$\frac{d\epsilon_N}{dt} = \frac{\pi(v/c)}{2\gamma}\langle\beta\rangle\left(\frac{15\ \text{MeV}}{mv^2}\right)^2 \frac{c}{L_\text{rad}} \tag{7.133}$$

$$= \pi\langle\beta\rangle(1.6 \times 10^{-7}/\text{sec})\frac{P[\mu\text{Torr}]}{\gamma} \tag{7.134}$$

for the assumptions above.

7.3 EMITTANCE REDUCTION

The processes described in the last two sections are indicative of common and often unavoidable sources of emittance dilution in hadron synchrotrons. Potential major sources of emittance growth can often be identified, and modifications can be made during the design of the accelerator to improve the situation. One may still have residual effects, which may be difficult to identify, and one may still desire smaller emittance beams than can be readily produced by normal means. In particular, beams of exotic particles, such as antiprotons, which are produced through the targeting of primary beams will have inherently large transverse emittances as well as large energy spread. It was the desire to build proton-antiproton (p$\bar{\text{p}}$) colliding beam accelerators which led to the development of electron cooling and stochastic cooling. These emittance reduction techniques, in particular the latter, allowed antiprotons to be accumulated in reasonable quantities and with reasonable emittances so that p$\bar{\text{p}}$ collisions in the 0.5–2 TeV center of mass energy range could be achieved. The invention of stochastic beam cooling and the subsequent discovery of the vector bosons at CERN in the Sp$\bar{\text{p}}$S collider led to the award of the Nobel Prize in physics to C. Rubbia and S. van der Meer. Over the past several years beam cooling techniques have been utilized in several accelerators in high energy and nuclear physics experimental facilities.

Electron cooling[2] involves the thermal interplay of a proton beam and an electron beam toward equilibrium transverse and longitudinal temperatures; the initial electron temperature is quite low (parallel beam with low momentum spread), and the initial proton beam temperature is significantly higher. While this technique is qualitatively simple to visualize, a complete quantitative treatment is beyond the scope of this section. Rather, we turn to a description of stochastic cooling. The basis of this technique is a shade less intuitive, yet its analysis is more straightforward.

7.3.1 Transverse Stochastic Cooling

The concept of a stochastic cooling system is remarkably simple. It is also remarkable that it is technically feasible.[3] Suppose we want to reduce the transverse emittance of a beam. A beam bunch contains a finite number of particles; thus the beam centroid will deviate from the central orbit of the bunch by a finite amount. If we detect and correct this deviation, the effective emittance of the bunch will be reduced. If we were indeed talking about a single bunch, we would have succeeded in making a minuscule reduction in the emittance and the process would be at an end. But if, on the other hand, our beam sample rapidly interchanges particles with other samples, the fluctuation in the centroid position is regenerated and the process can be repeated. In concept, at least, the system might consist of the arrangement in Figure 7.12. The centroid position is sensed at the pickup, and the signal is conveyed across the ring, amplified, and delivered to a kicker which provides an angular deflection proportional to the displacement sensed at the pickup. The kicker is located an odd number of quarter wavelengths in betatron phase downstream of the pickup. The signal path must be shorter than the

[2]G. I. Budker, *Proc. Intl. Symp. on Electron and Positron Storage Rings*, Saclay, 1966, p. II-1-1. For a general overview, see W. Kells, "Electron Cooling," in *Physics of High Energy Particle Accelerators*, AIP Conf. Proc. 87, New York, 1982.

[3]D. Möhl, G. Petrucci, L. Thorndahl, and S. van der Meer, "Physics and Technique of Stochastic Cooling," Physics Reports 58, No. 2 (1980). The following discussion has been adapted from A. V. Tollestrup and G. Dugan, "Elementary Stochastic Cooling," in *Physics of High Energy Particle Accelerators*, AIP Conf. Proc. 105, New York, 1983.

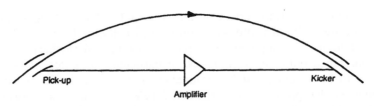

Figure 7.12. Stochastic cooling system consisting of pickup electrodes, amplifier, and beam deflector.

orbital path length between the two devices to ensure that the signal reaches the kicker at the same time as the beam.

It is desirable to reduce the transverse emittance of a beam containing a very large number of particles. If the system had infinitely fine time resolution, each particle could be sensed and corrected and the process could be successfully concluded in a few revolutions. However, a real system does not have this capability, and so there will be only finitely many particles in the sample that are corrected. Clearly, the smaller the number of particles in the sample, the closer one approaches the ideal.

First, we relate the sample size to the system bandwidth. Suppose there are N particles uniformly distributed around the ring. If the sample size is such that the beam is divided into k samples, each containing N_s particles, then the minimum wavelength that can be resolved in the analysis of the data is

$$\lambda_{\min} = \frac{2C}{k}, \tag{7.135}$$

where C is the ring circumference. Therefore, the frequency content of the information extends to

$$f_{\max} = \frac{v}{\lambda_{\min}} = \frac{kv}{2C} \approx \frac{k}{2T}, \tag{7.136}$$

where T is the revolution period. For a system with a flat frequency response from $f = 0$ to $f = W$, W determines $f_{\max}$. So the number of particles in a sample, in terms of the bandwidth W, is given by

$$N_s = \frac{N}{k} = \frac{N}{2TW}. \tag{7.137}$$

We now consider a measurement of a particular sample. Each particle in the sample receives a correction proportional to the sample's mean displacement $\langle x \rangle$. So an individual particle's displacement after the kick is $x - g\langle x \rangle$. To get at the emittance reduction, we need to consider the change of the rms of the distribution. For the kth particle,

$$x_k^2 \rightarrow \left(x_k - g\langle x \rangle\right)^2 = x_k^2 - 2gx_k\langle x \rangle + g^2\langle x \rangle^2. \tag{7.138}$$

We write

$$\langle x \rangle = \frac{1}{N_s}\sum_i x_i = \frac{1}{N_s}x_k + \frac{1}{N_s}\sum_{i \neq k} x_i. \tag{7.139}$$

Then

$$\left(x_k - g\langle x \rangle\right)^2 = x_k^2 - \frac{2g}{N_s}x_k^2 - \frac{2g}{N_s}x_k\sum_{i \neq k} x_i + \frac{g^2}{N_s^2}\left(x_k + \sum_{i \neq k} x_i\right)^2$$

$$= x_k^2 - \left(\frac{2g}{N_s} - \frac{g^2}{N_s^2}\right)x_k^2$$

$$- \left(\frac{2g}{N_s} - \frac{2g^2}{N_s^2}\right)\sum_{i \neq k} x_i x_k + \frac{g^2}{N_s^2}\left(\sum_{i \neq k} x_i\right)^2. \tag{7.140}$$

Averaging over all the particles, we get

$$\frac{1}{N_s}\sum_k \left(x_k - g\langle x \rangle\right)^2 = \frac{1}{N_s}\sum_k x_k^2 - \left(\frac{2g}{N_s} - \frac{g^2}{N_s^2}\right)\frac{1}{N_s}\sum_k x_k^2$$

$$- \left(\frac{2g}{N_s} - \frac{2g^2}{N_s^2}\right)\sum_{k, i \neq k} x_i x_k$$

$$+ \frac{g^2}{N_s}\sum_k \left(\frac{1}{N_s}\sum_{i \neq k} x_i\right)^2. \tag{7.141}$$

The second term on the right hand side is the sum of the contribution of each particle acting back upon itself. This coherent term is offset in part by the fourth term, representing the incoherent contribution of the other particles in the sample. Since the individual particle displacements are uncorrelated, the sum present in the third term is zero.

We can analyze the last term as follows:

$$\sum_k \left(\frac{1}{N_s} \sum_{i \neq k} x_i \right)^2 \approx \sum_k \left(\frac{1}{N_s} \sum_i x_i \right)^2$$

$$= \sum_k \frac{1}{N_s^2} \sum_i \sum_j x_i x_j$$

$$= \frac{1}{N_s} \sum_k \frac{1}{N_s} \sum_i x_i^2$$

$$= \frac{1}{N_s} \sum_k \langle x^2 \rangle$$

$$= \frac{1}{N_s} \cdot N_s \langle x^2 \rangle$$

$$= \langle x^2 \rangle, \qquad (7.142)$$

where in the second step we have assumed that the various x_i are uncorrelated. Thus, keeping terms up to first order in $1/N_s$, we have for the rate of change of $\langle x^2 \rangle$

$$\frac{d\langle x^2 \rangle}{dn} = -\frac{2g}{N_s} \langle x^2 \rangle + \frac{g^2}{N_s} \langle x^2 \rangle. \qquad (7.143)$$

The cooling rate is then

$$\frac{1}{\epsilon} \frac{d\epsilon}{dn} = -\left(\frac{2g - g^2}{N_s} \right), \qquad (7.144)$$

or, in terms of time,

$$\frac{1}{\tau} \equiv -\frac{1}{\epsilon} \frac{d\epsilon}{dt} = -\frac{1}{\epsilon} \frac{d\epsilon}{dn} \frac{1}{T} = \frac{2g - g^2}{N_s T} = \frac{2W}{N} (2g - g^2). \qquad (7.145)$$

Let's add two refinements to this relationship. System noise is an important consideration in the design of a cooling ring. Suppose that the noise introduced at the kicker is equivalent to a position error x_n at the pickup. Then the correction applied to each particle becomes

$$x - g(\langle x \rangle + x_n). \qquad (7.146)$$

Proceeding as before,

$$[x - g(\langle x \rangle + x_n)]^2 = x^2 - 2gx(\langle x \rangle + x_n) + g^2(\langle x \rangle^2 + 2x_n\langle x \rangle + x_n^2)$$
(7.147)

for a single particle, and averaging over the sample gives

$$\langle [x - g(\langle x \rangle + x_n)]^2 \rangle = \langle x^2 \rangle - 2g\langle x \rangle^2 - 2g\langle x \rangle\langle x_n \rangle + g^2\langle x \rangle^2$$
$$+ g^2\langle x_n \rangle\langle x \rangle + g^2\langle x_n^2 \rangle.$$
(7.148)

Averaging over many samples, $\langle x_n \rangle = 0$ and so

$$\frac{1}{\langle x^2 \rangle} \frac{d\langle x^2 \rangle}{dn} = [-2g + g^2(1 + U)]\frac{1}{N_s},$$
(7.149)

where $U \equiv \langle x_n^2 \rangle/\langle x \rangle^2$ is the ratio of the expected noise to the expected signal power.

Our second refinement is to take into account the fact that the fluctuation in the centroid position may not be regenerated independently from one turn to the next. In other words, if particles move rapidly from one sample to another, each sample will rerandomize during the course of one turn and we will have the ideal situation. But the "mixing" may not be perfect, and we have to allow for this possibility.

The movement from sample to sample is due to the spread in orbital frequencies arising from the spread in particle momentum. The number of revolutions required for a particle of momentum $\Delta p/p$ to pass from one sample to another is

$$M = \frac{T_s}{\Delta T},$$
(7.150)

where $T_s = (N_s/N)T = 1/(2W)$ is the sample time, and ΔT is the change in the revolution period due to the momentum deviation $\Delta p/p$. Then

$$M = \frac{1}{2WT|\eta|(\Delta p/p)}.$$
(7.151)

For ideal mixing, $M = 1$. Intuitively, one would expect the cooling rate to degrade by a factor of M as we depart from perfect mixing. Actually, this factor of M appears only in the incoherent term, and so the emittance decreases according to

$$\epsilon = \epsilon_0 e^{-t/\tau},$$
(7.152)

where we have for the cooling rate

$$\frac{1}{\tau} = \frac{2W}{N}\left[2g - g^2(M + U)\right]. \tag{7.153}$$

7.3.2 Longitudinal Stochastic Cooling

The transverse cooling sketched in the last section is able to reduce the transverse emittances of antiprotons to the level appropriate for $p\bar{p}$ collider operation. We have not thus far addressed the question of how one accumulates large numbers of antiprotons. It is inherent in the production process that antiprotons are produced over a broad range of momenta, and this spread needs to be reduced. Longitudinal cooling is able to achieve both goals, as we shall see.

Suppose we detect momentum differences by their related orbital frequency differences. We need a way of applying no correction if the frequency is correct; this can be accomplished by adding a filter to the layout shown in the preceding section. If a correction is required, it will be applied by a longitudinal kick, rather than a transverse kick as was done in the previous section. Because the cooling systems have a wide bandwidth, this implies that the filter remove not only the fundamental of the derived frequency but its harmonics as well.

In a system devised for accumulation of particles it is natural to speak in terms of particle flux and density functions. The time evolution of the density function $\psi(E)$ will represent a trade off between the diffusive effects of the incoherent interactions and the collective flow arising from the coherent forces. The equation that describes the time evolution of a density function subject to these processes is called the Fokker–Planck equation.

As in the discussion of beam-gas scattering, the flux arising from diffusion can be written in the form

$$\vec{J} = -D\,\nabla\psi, \tag{7.154}$$

where $\vec{J}$ is the particle flux. In the case under consideration here, since energy is the only degree of freedom,

$$J = -D(E)\frac{\partial\psi}{\partial E}, \tag{7.155}$$

where the diffusion "constant" may be a function of energy. To this, we must add coherent forces. If the rate of energy gain is $C(E)$, then we must add $\psi C(E)$ to the flux, obtaining

$$J = C(E)\psi - D(E)\frac{\partial\psi}{\partial E}. \tag{7.156}$$

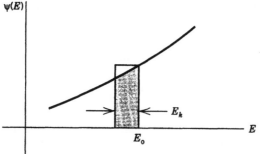

Figure 7.13. Particle density function $\psi(E)$.

We can now obtain the time rate of change of ψ from the continuity equation:

$$\frac{\partial \psi}{\partial t} = -\nabla \cdot \vec{J} = -\frac{\partial}{\partial E}\left[C(E)\psi - D(E)\frac{\partial \psi}{\partial E}\right]. \qquad (7.157)$$

This is called the Fokker–Planck equation.

Let's arrive at the expression for the flux by an alternative route so that we may identify the coefficients $C(E)$ and $D(E)$ in terms of the kicker voltage. Suppose $\psi(E)$ appears as in Figure 7.13. We are interested in the flux at E_0 due to an energy increment E_k generated by the kicker. The number of particles present in the shaded area is

$$n = \psi(E_0)E_k - \tfrac{1}{2}E_k\left(\frac{\partial \psi}{\partial E}E_k\right). \qquad (7.158)$$

On the time scale associated with the kicker frequency, the particle distribution changes very little. Therefore, we may average over a time interval sufficiently short that ψ does not change, but sufficiently long to get meaningful averages of the kicker voltage factors. We get

$$\langle n \rangle = \psi(E_0)\langle E_k \rangle - \tfrac{1}{2}\langle E_k^2 \rangle\left(\frac{\partial \psi}{\partial E}\right)_{E_0}. \qquad (7.159)$$

Therefore, the average flux of particles passing through this region will be

$$J = \frac{d\langle n \rangle}{dt} = \psi\frac{d}{dt}\langle E_k \rangle - \frac{1}{2}\frac{d}{dt}\langle E_k^2 \rangle\frac{\partial \psi}{\partial E}. \qquad (7.160)$$

By comparison with our earlier expression for the flux, the coherent force coefficient and the diffusion (noise) coefficient are given by

$$C(E) = \frac{d}{dt}\langle E_k \rangle \tag{7.161}$$

$$D(E) = \frac{1}{2}\frac{d}{dt}\langle E_k^2 \rangle. \tag{7.162}$$

We may now apply the above relationships to momentum stacking and cooling. We will examine two cases for which the particle density does not depend upon time. In such equilibrium circumstances the Fokker–Planck equation tells us that the flux is constant. We first consider a simple example in which the flux is zero. Suppose there is a coherent force driving particles toward some central energy E_0, where the force is proportional to the energy deviation $E - E_0$, and suppose that the diffusion force is a constant. Then $C(E) = -\alpha(E - E_0)$ and $D(E) = D_0$. So this static situation is described by

$$J = -\alpha(E - E_0)\psi - D_0\frac{\partial \psi}{\partial E} = 0. \tag{7.163}$$

The solution to the above equation is the Gaussian

$$\psi = \psi_0 e^{-\alpha(E-E_0)2/2D_0}. \tag{7.164}$$

A large particle density results if the noise (D_0) is small and if the restoring force (α) is large.

Now suppose we were to introduce a small group of particles with central energy E_i, where $E_i - E_0$ is large compared to the rms of the distribution above. Then the coherent force would dominate the force due to diffusion, and this small group would be driven toward the larger distribution over some time interval, as depicted in Figure 7.14. Such a scenario is referred to as momentum stacking; small pulses of particles are continuously injected into the synchrotron at an energy E_i and then are collected into an equilibrium distribution with a central core at E_0.

The remarks concerning momentum stacking in the preceding paragraph suggest the basis for the method of antiproton accumulation used at CERN and Fermilab. In this method—the Van der Meer method—the flux is constant with time, with particles continuously being injected into the accumulator storage ring. We note that the coherent force, in terms of the voltage $V(E)$ applied by the kicker each turn, would be

$$C(E) = \frac{eV(E)}{T} \tag{7.165}$$

where T is the revolution period. To arrive at an approximate expression for

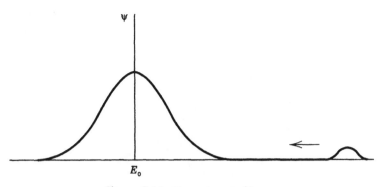

Figure 7.14. Momentum stacking.

the diffusion coefficient, assume that the expectation value of E_k^2 arises solely from the incoherent noise in the sample. Then, recalling the argument in the preceding section relating $\langle x^2 \rangle$ to $\langle x \rangle$,

$$\langle E_k^2 \rangle = \langle E_k \rangle^2 \times N_s. \tag{7.166}$$

In this case, since we are sampling frequencies (and therefore energies), the number of particles in the sample is proportional to ψ. Hence, we expect the diffusion coefficient to be of the form

$$D(E) = AV^2\psi, \tag{7.167}$$

where A is a constant determined by the design of the cooling system.

So, putting this all together, the constant flux is given by

$$J = \frac{eV}{T}\psi - AV^2\psi \frac{\partial \psi}{\partial E} = J_0. \tag{7.168}$$

Solving for $\partial\psi/\partial E$, we have

$$\frac{\partial \psi}{\partial E} = -\frac{J_0}{AV^2\psi} + \frac{e}{AVT}. \tag{7.169}$$

We now choose a kicker voltage which will make $\partial\psi/\partial E$ as large as possible:

$$V = \frac{2TJ_0}{e\psi}. \tag{7.170}$$

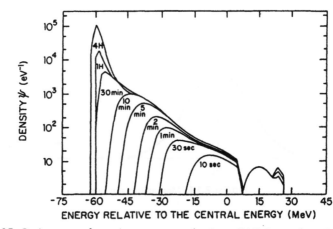

Figure 7.15. Design curves for antiproton energy density at FNAL Accumulator Ring, showing development of the "p̄ stack" over time. From Tollestrup and Dugan, with permission.

So we have

$$\frac{\partial \psi}{\partial E} = -\frac{e^2 \psi}{4 J_0 T^2 A} + \frac{e^2 \psi}{2 J_0 T^2 A} = \frac{e^2 \psi}{4 A T^2 J_0} \equiv \frac{\psi}{E_d}, \qquad (7.171)$$

or

$$\psi = \psi_0 e^{(E - E_i)/E_d}. \qquad (7.172)$$

Therefore, in the equilibrium state, there is a constant flux of particles being injected at energy E_i; over time, the particle density increases exponentially with the energy difference $E - E_i$ as shown in Figure 7.15. [In this figure, the particle flux is negative (from the right) and hence the density increases to the left as shown.]

To generate the density profile described above, the kicker voltage must be given by

$$V = \frac{2 T J_0}{e \psi} = \frac{2 T J_0}{e \psi_0} e^{-(E - E_i)/E_d}, \qquad (7.173)$$

which tells us that the particles in the higher density region need to receive less kick in an exponential fashion. As can be seen in Figure 7.15, the particle density in the core can be increased many orders of magnitude in just a few hours. For a sense of scale, the central energy of this accumulator ring is 8 GeV.

7.4 SOME REMARKS ON BEAM DISTRIBUTIONS

We have often, throughout the text, assumed that particle distributions are Gaussian in order to perform calculations. The question always arises whether or not this is a reasonable assumption. It is easy to conceive of ways of producing beam distributions which are not Gaussian, such as those produced in a hadron storage ring by deflecting the beam with a pulsed kicker magnet and allowing filamentation to occur; indeed, this experiment has been

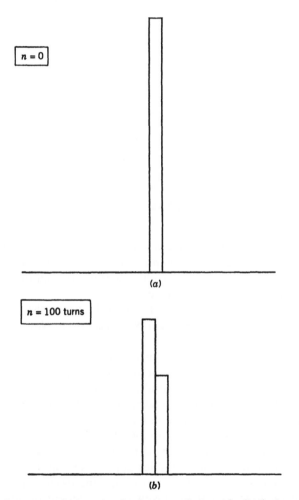

Figure 7.16. Histograms showing the development of a particle distribution due to random changes in particle trajectories. The initial particle distribution function is given by $n(x) = N\delta(0)$, where N is the total number of particles (200 for the case shown) and $\delta(x)$ is the Dirac δ-function. Independent of the choice of the random process, the resulting distribution will appear Gaussian after enough time has elapsed.

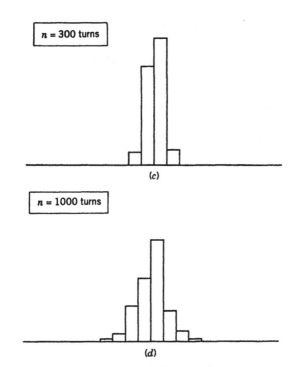

Figure 7.16. (*Continued*).

performed many times. But if care is taken to match trajectories and optical properties at injection and to minimize other coherent disturbances to the beam, then abrupt changes to individual particle trajectories are apt to be dominated by purely random processes.

Consider, for example, a zero emittance initial beam distribution—that is, all the particles start out at the origin of phase space. Now suppose that on each turn about the accelerator each particle receives a random change in the slope of its trajectory—either a positive increment of one unit, a negative increment of one unit, or no change at all. Each particle will undergo its own independent random walk. The results of such a scenario are shown in Figure 7.16. Over a large number of such increments and for a large number of particles, the distribution of transverse positions will tend to have the shape of the normal (Gaussian) distribution function. This is just the manifestation of the central limit theorem of probability theory:[4] the distribution function for a sum of random variables approaches the normal distribution function as the number of variables in the sum increases. The most powerful aspect of the theorem is that the result is independent of how the fluctuations are distributed, just so long as they are random.

---

[4]See, for example, H. J. Larson, *Introduction to Probability Theory and Statistical Inference*, John Wiley & Sons, New York, 1974.

Hence, one can imagine a beam of protons, each particle having suffered various random deflections due to scattering from gas molecules, intrabeam scattering, perturbations in magnetic fields, power supply noise, mechanical vibrations, and so on. So long as the various events are random and uncorrelated among the various particles, the distribution function describing the position of a particle which has been subjected to all of these random processes will be Gaussian, to good approximation.

At the beginning of this chapter it was mentioned that discussion of electron beam emittance would be left to Chapter 8. There we will see that the dominant source of emittance fluctuation is the emission of photons due to synchrotron radiation, an inherently random process. Since the ensuing damping times are, in general, short, one would expect that even correlated processes such as injection errors might be swamped by the synchrotron radiation effects. Thus, electron beams in circular accelerators are certainly expected to have Gaussian distributions, and that is what is observed. The fact that proton beams in large synchrotrons also typically appear Gaussian suggests that the central limit theorem is at work here as well.

PROBLEMS

1. Compute the increase in the normalized (39%) emittance due to a 1 mm amplitude steering error observed at a point where $\beta = 100$ m for injection energies of (a) 8 GeV, (b) 150 GeV, and (c) 2 TeV.

2. Suppose an injection line leading into a planar synchrotron leads to a nonzero value for the vertical dispersion function. Show that the variance of the resulting vertical distribution after dilution takes place is related to the variance of the incoming distribution by

$$\sigma^2 = \sigma_0^2 + \frac{1}{2}\left(\Delta D \frac{\sigma_p}{p}\right)^2,$$

where σ_p/p is the rms of the distribution in $\Delta p/p$, and ΔD is the value of the dispersion function delivered by the beamline to the observation point. The slope of the dispersion function is assumed to be well matched.

3. Show that the quantity F used in the discussion of amplitude function mismatch may be written as

$$F = 1 - \tfrac{1}{2}\det(J - J_0),$$

where the J's are the matrices of Courant-Snyder parameters used in the text. Here, J_0 reflects the values of the synchrotron lattice, while J contains the parameters delivered by the beamline.

4. Using the result of the previous problem, show that a quadrupole located in a transfer line between two accelerators which has a field error of $\Delta B'/B'$ will produce an emittance dilution given by

$$\frac{\epsilon}{\epsilon_0} = 1 + \frac{1}{2}\left(\frac{\beta_0}{f}\right)^2\left(\frac{\Delta B'}{B'}\right)^2,$$

where f is the nominal focal length of the quadrupole and β_0 is the design value of the amplitude function at the location of the quadrupole.

5. Show that $\det(\Delta J)$ is an invariant within an unperturbed lattice, independent of longitudinal coordinate s. Thus, the choice of an injection point is arbitrary when discussing the mismatch of Courant-Snyder parameters.

6. Show that the emittance dilution factor due to an amplitude function mismatch can be written as

$$F = 1 + \frac{1}{2}\left[\frac{(\Delta\beta/\beta_0)^2 + (\alpha_0(\Delta\beta/\beta_0) - \Delta\alpha)^2}{1 + (\Delta\beta/\beta_0)}\right]$$

7. Downstream of a gradient perturbation, show that the amplitude function mismatch propagates through the unperturbed lattice of a synchrotron obeying the differential equation

$$\frac{d^2}{d\phi^2}\left(\frac{\Delta\beta}{\beta_0}\right) + 4\nu_0^2\left(\frac{\Delta\beta}{\beta_0}\right) = -2\nu_0^2\det(\Delta J)$$

where $\phi \equiv \psi/\nu_0$ is the reduced phase and ν_0 is the unperturbed tune.

8. Suppose a synchronous transfer occurs between two accelerators which have their RF frequencies and bucket areas properly matched in order to preserve longitudinal emittance in the transfer. If the bunch area in phase space is much smaller than the bucket area, compute the change in longitudinal emittance due to a small (a) energy mismatch, (b) phase mismatch.

9. In the text, we discuss how injection errors lead to an increase in emittance. Liouville's theorem states that phase space density is conserved for Hamiltonian systems. Explain this apparent contradiction.

10. Consider the $x, (\beta x' + \alpha x)$ phase space distribution generated by a steering error of amplitude Δx inflicted upon a Gaussian beam of initial rms size σ_0. Compute the radius a_0 of the phase space circle which contains 39% of the injected particles. Define a dilution factor a_0^2/σ_0^2, and compare this with the expression $\sigma^2/\sigma_0^2 = 1 + \frac{1}{2}\Delta x^2$ found in the text.

11. Verify that

$$f(Z,\tau) = \sum_n c_n J_0\left(\lambda_n \sqrt{Z}\right) e^{-\lambda_n^2 \tau/4}$$

is indeed the appropriate solution of the diffusion equation, as was stated in Equation 7.77.

12. Suppose the normalized emittance of the beam injected into the Fermilab Main Ring is 15π mm mrad (95%). Assume that the average vacuum pressure is 5×10^{-7} Torr. If the limiting half aperture of the accelerator is 10 mm at a location of $\langle \beta \rangle = 50$ m, estimate the fractional beam loss due to scattering with the residual gas after 4 seconds for an injection energy of (a) 8 GeV and (b) 20 GeV. It may be helpful to make use of the graphs in the text.

13. Consider a beam which uniformly populates $x, (\beta x' + \alpha x)$ phase space out to a radius a_0. Show that the solution to the diffusion equation is

$$f(Z,\tau) = \sum_n \frac{2 J_1(\lambda_n a_0/a)}{\lambda_n J_1^2(\lambda_n)} \frac{a}{a_0} J_0\left(\lambda_n \sqrt{Z}\right) e^{-\lambda_n^2 \tau/4},$$

$$N(\tau) = 4 \frac{a}{a_0} \sum_n \frac{J_1(\lambda_n a_0/a)}{\lambda_n^2 J_1(\lambda_n)} e^{-\lambda_n^2 \tau/4},$$

where a is the limiting aperture.

14. It is now common to inject negative hydrogen ions delivered by a linear accelerator into the first synchrotron of a large proton accelerator facility. The electrons are stripped by passing the incoming beam through a carbon foil. Estimate the emittance increase if the particles pass through a foil of 25 μm thickness every 1.6 μsec for a total of 16 μsec.

15. Suppose the luminosity lifetime

$$\frac{1}{\tau} \equiv -\frac{1}{\mathcal{L}} \frac{d\mathcal{L}}{dt}$$

in the Tevatron collider is 10 hours, and suppose it is due to transverse emittance growth of both particle species. If the emittance is attributed to RF noise, estimate the rms noise voltage. Assume $\epsilon_N(95\%) \approx 20\pi$ mm mrad. The horizontal lattice functions at the accelerating stations are $\beta = 72$ m, $\alpha = -0.47$, $D = 2.4$ m, and $D' = 0.02$. Comment on the role of gas scattering in the luminosity lifetime.

16. Transverse betatron cooling in the Fermilab Debuncher ring is carried out with a 2–4 GHz system. The orbit period is 1.6 μsec, the slip factor η is 0.006, the momentum spread may be characterized as 0.3%, and the initial noise-to-signal ratio is ~ 2. Calculate the optimum cooling rate for a beam of 10^7 antiprotons.

17. The appeal of electron cooling is easy to illustrate. In conventional kinetic theory, the gas temperature is related to the mean energy of the molecules by

$$\frac{\langle p^2 \rangle}{2m} = \frac{3}{2}kT.$$

So for an ion beam, one can define a "temperature" for each degree of freedom by

$$\frac{\langle p_{x0}^2 \rangle}{m}, \quad \frac{\langle p_{y0}^2 \rangle}{m}, \quad \frac{\langle p_{s0}^2 \rangle}{m},$$

where the Boltzmann constant has been supressed. The subscript "0" implies that the momenta are measured with respect to the rest frame of the beam centroid.

(a) Show that for one transverse degree of freedom

$$T_x = mc^2 \left(\frac{v}{c} \right)^2 \gamma^2 (\sigma')^2 = mc^2 \left(\frac{v}{c} \right) \gamma \frac{\epsilon_N}{\pi \beta},$$

where all the quantities are now measured in the laboratory frame. Here, $(\sigma')^2 = \langle (x')^2 \rangle$. Note that, because of the presence of the amplitude function β, the temperature is a function of position.

(b) Evaluate T_x for typical injection parameters from a proton linac into a synchrotron. Take $(v/c)\gamma \approx 0.7$, $\epsilon_N = \pi/2$ mm mrad, and $\beta = 10$ m.

(c) Repeat the calculation for the longitudinal degree of freedom. Show that

$$T_s = mc^2 \left(\frac{v}{c} \right)^2 \sigma_p^2, \quad \sigma_p^2 \equiv \left\langle \left(\frac{\Delta p}{p} \right)^2 \right\rangle.$$

Take $\sigma_p = 10^{-3}$ to obtain a numerical estimate.

(d) In electron cooling, an electron beam traveling at the same speed as the ion beam centroid interchanges energy with the ion beam. Estimate the temperature of an electron beam emitted from a hot cathode, in the same units as that used for the ion temperatures above.

18. In this chapter, we have concentrated on transverse emittance growth, in large part because of the importance of transverse emittance to the luminosity of a collider. But longitudinal emittance cannot be ignored, for eventually dilution processes may lead to loss of particles from stable buckets. Derive an expression for longitudinal emittance growth analogous to Equation 7.112.

CHAPTER 8

Synchrotron Radiation

Synchrotron radiation radiation is the dominant factor in the design of high energy electron synchrotrons and is the obstacle to exceeding 100 GeV or so in this type of accelerator. It has also brought about the spectacular success of synchrotron light sources. Only today is synchrotron radiation becoming a design consideration for proton synchrotrons. In the proton case, single particle motion, to a very good approximation, exemplifies a Hamiltonian system. Particle motion in electron synchrotrons, on the other hand, is inherently dissipative.

In this chapter, we look at some of the basic properties of the radiation process, the power, and the characteristic photon energy. The energy loss due to synchrotron radiation and its replacement by the RF acceleration system leads to a variation with time of the oscillation amplitudes in all three degrees of freedom, and we compute the related time constants. We will reproduce an elegant theorem—Robinson's theorem—which relates these three time constants and demonstrates that there is a net damping effect that can be apportioned among the degrees of freedom at the choice of the designer. An electron storage ring will be designed to damp in each degree of freedom.

The fact that the radiation process is quantized implies that there are statistical fluctuations in the radiation rate. These fluctuations cause excitation of synchrotron oscillations, and of betatron oscillations in at least one transverse degree of freedom. The interplay between the quantum fluctuations and damping will result for an ensemble of particles in an equilibrium beam distribution, which we will find to be Gaussian.

8.1 RADIATION FROM RELATIVISTIC PARTICLES

If a slowly moving particle of charge e undergoes an acceleration a, then the radiated power P is given by the Larmor formula:

$$P = \frac{1}{6\pi\epsilon_0}\frac{e^2 a^2}{c^3}.\tag{8.1}$$

The angular distribution of the radiation varies as $\sin^2\theta$, where θ is the angle between the direction of the acceleration and the point of observation.

We can find the radiated power for relativistic charges by using the fact that radiated power is a Lorentz invariant. To arrive at the latter conclusion, we can argue as follows. Suppose a photon of angular frequency ω' is traveling at a direction θ with respect to the x'-axis in a "primed" frame that is moving parallel to the x-axis of the "unprimed" laboratory frame. Transformation to the unprimed frame gives

$$\tan\theta = \frac{\sin\theta'}{\gamma(\cos\theta' + \beta)},\tag{8.2}$$

$$\omega = \gamma\omega'(1 + \beta\cos\theta'),\tag{8.3}$$

where γ is the Lorentz factor characterizing the relative motion of the two frames. If two photons are emitted at angles θ' and $\theta' + \pi$ with the same angular frequency ω', then in the laboratory frame the total energy will be proportional to

$$\omega_1 + \omega_2 = 2\gamma\omega'.\tag{8.4}$$

If the emission takes place in a short interval τ', then in terms of radiated power, the relation above can be written

$$P\tau = P'\tau'\gamma,\tag{8.5}$$

or $P = P'$, after recognition of the effect of time dilation. That is, the power that is lost to the Doppler shift in one direction is gained back in the other. So long as the angular distribution of radiation in the primed frame has the appropriate symmetry, we can conclude that the power is an invariant. Though not developed in all generality, this is enough for our purposes here.

Look at two cases: acceleration perpendicular to and parallel to the direction of motion of a relativistic charge. The first corresponds to a particle undergoing deflection in a bending magnet. In an inertial frame traveling at the speed of the particle and tangent to the orbit at the time of arrival of the particle, at that instant the particle will be at rest and undergoing acceleration in the $-y'$ direction. This situation is shown in Figure 8.1. In this

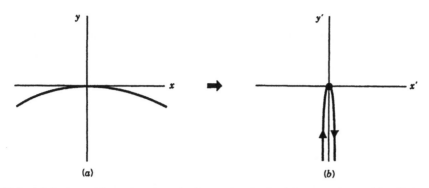

Figure 8.1. A particle undergoing circular motion in the laboratory frame (a) will come momentarily to rest when viewed in a frame moving tangent to its trajectory (b).

(primed) frame the power radiated is given by the Larmor expression, with a' inserted for the acceleration. For acceleration transverse to the relative direction of motion of the two frames, $a' = \gamma^2 a$. In the primed frame, the power distribution has the proper front-back symmetry needed for the argument of the preceding paragraph to be valid, so the power is an invariant. As a result, the power in the laboratory frame is

$$P = \frac{1}{6\pi\epsilon_0} \cdot \frac{e^2 a^2}{c^3} \gamma^4$$

$$= \frac{1}{6\pi\epsilon_0} \frac{e^2 c}{\rho^2} \gamma^4$$

$$= \frac{1}{6\pi\epsilon_0} \frac{e^4}{m^4 c^5} B^2 E^2. \tag{8.6}$$

In the second form, a has been replaced by the centripetal acceleration c^2/ρ of a relativistic electron. The third form will be useful when we discuss radiation damping; here, E is the total energy, $E = \gamma mc^2$, and B is the magnetic field producing the curvature of the particle's path.

In contrast, suppose that the acceleration is in the direction of motion of the charge. In the primed frame, the angular distribution just rotates by $\pi/2$, so our invariance-of-power argument is still all right. But now $a' = \gamma^3 a$, and so the power radiated in the laboratory frame is

$$P = \frac{1}{6\pi\epsilon_0} \frac{e^2 a^2}{c^3} \gamma^6. \tag{8.7}$$

At first glance, this result looks even more ominous than the one for transverse acceleration. But such is not the case—acceleration in the direction of motion of a rapidly moving particle is not as easily produced as transverse acceleration. Equation 8.7 can be recast in the form

$$P = \frac{2}{3} \frac{r_0}{mc} \dot{p}^2,\qquad(8.8)$$

where r_0 is the classical radius of the particle. Take one of the factors of $\dot{p}$, express it in terms of the rate of change of energy of the particle, $\dot{E}$, and compare P with $\dot{E}$. Write the other factor of $\dot{p}$ just as the force F. Then we have

$$\frac{P}{\dot{E}} = \frac{2}{3} \frac{r_0 F}{mc^2 \beta},\qquad(8.9)$$

and for the ratio on the left to be significant, the particle must experience an energy gain or loss comparable with its rest energy within a distance equal to its classical radius. That is possible on the atomic or nuclear scale (e.g., bremsstrahlung), but not with laboratory sized accelerator components.

So we will be concerned only with radiation arising from transverse acceleration. The radiation loss per turn on the design orbit of a synchrotron will be

$$U_0 = \int_0^{2\pi R} P\,dz/c\qquad(8.10)$$

$$= C_\gamma E^4 R \left\langle \frac{1}{\rho^2} \right\rangle,\qquad(8.11)$$

where

$$C_\gamma = \frac{4\pi}{3} \frac{r_0}{(mc^2)^3} = 8.85 \times 10^{-5} \frac{\text{meters}}{\text{GeV}^3},\qquad(8.12)$$

and the square of the curvature $1/\rho$ is averaged over the circumference $2\pi R$ of the ring. The numerical coefficient has been evaluated for the electron, using $r_0 = 2.818 \times 10^{-15}$ m. The average power radiated is

$$\langle P \rangle = fU_0,\qquad(8.13)$$

where f is the orbit frequency $c/2\pi R$.

8.2 DAMPING OF OSCILLATIONS

With the inclusion of synchrotron radiation, the transverse and longitudinal oscillations of a single particle no longer have invariant amplitudes, for the system is now dissipative. In this section we calculate the damping rates. Note that a characteristic time for synchrotron radiation effects is the time τ_0 in which an electron of energy E would radiate E, i.e.,

$$\tau_0 \equiv \frac{E}{\langle P \rangle}, \tag{8.14}$$

and the rates will be expressed in terms of τ_0.

Damping of vertical betatron oscillations is easy to understand. Synchrotron radiation reduces the momentum of a particle in the direction of its motion, while the acceleration system restores momentum parallel to the central orbit. Consider the case in which there is no net acceleration, as in beam storage. On the average, the two momentum increments are equal in magnitude. If, in an element of path ds, the particle radiates energy du and receives the same energy increment from the acceleration system, then the momenta before and after, $\vec{p}_1$ and $\vec{p}_2$, are related by

$$\vec{p}_2 = \vec{p}_1 - \frac{du}{c} \frac{\vec{p}_1}{|\vec{p}_1|} + \frac{du}{c} \hat{s}. \tag{8.15}$$

as seen in Figure 8.2. In terms of the transverse and longitudinal components,

$$p_{2y} = p_{1y} - \frac{du}{c} \frac{p_{1y}}{|\vec{p}_1|}, \tag{8.16}$$

$$p_{2s} = p_{1s} - \frac{du}{c} \frac{p_{1s}}{|\vec{p}_1|} + \frac{du}{c} \tag{8.17}$$

Division of the first by the second gives the relationship between $y' = p_y/p_s$ before and after traversing ds:

$$y'_2 = y'_1 \frac{1 - du/E}{1 - du/E + du/(cp_s)}$$

$$= y'_1 (1 - du/E), \tag{8.18}$$

where E is the total energy of the particle, and we have kept only the lowest order term in du/E in the second equation. Therefore, the equation of

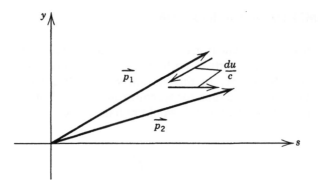

Figure 8.2. A particle undergoing a vertical betatron oscillation radiates in its direction of motion, and the momentum is restored in the direction of the design trajectory.

motion in y contains a term

$$y'' = -\frac{1}{E}\frac{du}{ds}y' \tag{8.19}$$

in addition to the focusing term proportional to y. For a damping rate slow compared to the betatron oscillation frequency, the free oscillation then is just modified by the multiplicative factor

$$\exp\left(-\frac{1}{2}\int\frac{1}{E}\frac{du}{ds}\,ds\right) = \exp\left(-\frac{t}{2\tau_0}\right), \tag{8.20}$$

and so the damping time constant is

$$\tau_y = 2\tau_0. \tag{8.21}$$

Identification of the damping time constant for synchrotron oscillations is almost as easy. Suppose the deviation from the synchronous energy of a particle is ΔE. In traversing an infinitesimal element of a turn, ΔE will change according to

$$\Delta E_2 = \Delta E_1 - du(\Delta E_1) + du(0), \tag{8.22}$$

where the second and third terms on the right are the energy loss due to synchrotron radiation at energy displaced from the synchronous energy ΔE_1 and the energy gain from the radiofrequency system at $\Delta E = 0$ respectively.

In terms of the radiated power,

$$du(\Delta E_1) = P(\Delta E_1)dt_1$$

$$= P(0)\left[1 + 2\frac{\Delta E_1}{E} + 2\frac{\Delta B}{B}\right]\left[1 + \frac{D}{\rho}\frac{\Delta E_1}{E}\right]dt_0, \quad (8.23)$$

$$du(0) = P(0)dt_0, \quad (8.24)$$

where dt_1 has been expressed in terms of the time element dt_0 on the synchronous orbit using Figure 3.16, and $P(\Delta E)$ has been written in terms of $P(0)$ using Equation 8.6. The ΔB in the first bracket of Equation 8.24 can be eliminated in favor of the fractional energy difference by the use of

$$\Delta B = B'x = B'D\frac{\Delta E}{E}. \quad (8.25)$$

Then to lowest order in the Δ's, the change in ΔE per turn due to synchrotron radiation becomes

$$\frac{d\Delta E}{dn} = -\Delta E\int_0^T \frac{P(0)}{E}\left[2 + \frac{D}{\rho} + 2D\frac{B'}{B}\right]dt_0. \quad (8.26)$$

Here, T is the period of the synchronous orbit, and as usual, we assume that changes in energy are small in that time scale. The first term in the integral is just $2U_0/E$, where, as before, U_0 is the energy radiated in one turn by the synchronous particle. If we change to time as the independent variable by multiplying both sides by the orbit frequency, f, we obtain

$$\frac{d\Delta E}{dt} = -\Delta E\left[2\frac{fU_0}{E} + \int_0^T \frac{fP(0)dt_0}{E}D\left(\frac{1}{\rho} + 2\frac{B'}{B}\right)\right]. \quad (8.27)$$

The quantity fU_0/E is $1/\tau_0$, where τ_0 is the characteristic time for radiation processes. If we take τ_0 outside of the brackets, then, after cancellation of various coefficients, the result is

$$\frac{d\Delta E}{dt} = -\frac{\Delta E}{\tau_0}(2 + \mathscr{D}), \quad (8.28)$$

$$\mathscr{D} \equiv \frac{\left\langle\frac{D}{\rho^2}\left(\frac{1}{\rho} + 2\frac{B'}{B}\right)\right\rangle}{\left\langle\frac{1}{\rho^2}\right\rangle}. \quad (8.29)$$

When this term is added to the equations of motion for a synchrotron oscillation, the solution for the motion will contain the factor

$$\exp\left[-\frac{1}{2\tau_0}(2 + \mathscr{D})\right],$$ (8.30)

and so the time constant for damping in this degree of freedom is

$$\tau_s = \frac{2\tau_0}{2 + \mathscr{D}}.$$ (8.31)

The analogous argument for the horizontal betatron oscillation is longer. It's more elegant to use *Robinson's theorem*, which deduces the sum of the damping rates for all three degrees of freedom.[1] Since we already know the results for two of the modes, the theorem gives us the third immediately.

The derivation goes as follows. Consider the transfer matrix of the six-vector $x, x', y, y', \phi, \Delta E$ through a path element ds. The diagonal elements for x' and y' will differ from unity by the quantity $-du/E$, as we have seen above. The diagonal element for ΔE will differ from unity by $-2\,du/E$. The only terms in the determinant of the matrix that are first order in ds come from the diagonal elements. So for this infinitesimal matrix

$$\det dM = 1 - 4\,du/E,$$ (8.32)

and since the determinant of a product of matrices is the product of their determinants, to lowest order for one revolution

$$\det M = 1 - 4\frac{U_0}{E}.$$ (8.33)

But the determinant is also the product of the eigenvalues. For oscillatory modes, the eigenvalues can be expressed as $\exp(\gamma_k)$. The six γ_k occur in conjugate pairs, so the imaginary parts do not contribute to the product. If we call the real parts α_x, α_y, and α_s, then

$$\alpha_x + \alpha_y + \alpha_s = -2\frac{U_0}{E}.$$ (8.34)

The α's are the decrements per turn; multiplication by the orbit frequency

[1]K. W. Robinson, "Radiation Effects in Circular Electron Accelerators," Phys. Rev. **111**, No. 2 (1958).

gives the result for the time constants:

$$\frac{1}{\tau_x} + \frac{1}{\tau_y} + \frac{1}{\tau_s} = \frac{2}{\tau_0}.$$

(8.35)

Therefore, using Robinson's theorem, we see that the damping time in the horizontal degree of freedom is

$$\tau_x = \frac{2}{1 - \mathscr{D}}\tau_0.$$

(8.36)

For a separated function synchrotron, where the focusing and bending are performed by separate elements, then $\mathscr{D}$ is small, and so for this case $\tau_x = 2\tau_0$ and all three degrees of freedom damp. This is, of course, the behavior one wishes for a storage ring.

8.3   QUANTUM FLUCTUATIONS AND EQUILIBRIUM BEAM SIZE

If the results of the preceding section were the end of the subject, we could design an electron storage ring in which all three degrees of freedom damped, and the emittances would shrink to zero. But such is not the case. The radiation process proceeds through the emission of discrete quanta, and the fluctuations in this random process produce an excitation of horizontal betatron oscillations and synchrotron oscillations.

To see how this excitation comes about, suppose that a particle is traveling along its synchronous orbit and emits a photon of energy w. The position of the particle doesn't change, so it suddenly finds itself starting a synchrotron oscillation with an initial energy offset $-w$ and a horizontal betatron oscillation with initial conditions $x = Dw/E$ and $x' = D'w/E$. Because of the random character of the photon emission, synchrotron radiation contributes a constant term to the growth of the horizontal and longitudinal emittances. This is just the situation we encountered in Chapter 7 when discussing emittance growth due to RF noise. Using that result, we expect the variance of the horizontal particle distribution to increase at the rate

$$\frac{d\sigma_x^2}{dt} = \tfrac{1}{2}Nf_0\langle\mathscr{H}\rangle\beta_x\frac{\langle w^2\rangle}{E^2},$$

(8.37)

where $\mathscr{H}$ is defined in Equation 7.106, β_x is the amplitude function, and N is the number of photon emissions per turn:

$$N = \frac{\langle P\rangle}{f_0\langle w\rangle} = \frac{U_0}{\langle w\rangle}.$$

(8.38)

In an ideal planar synchrotron, where there is no vertical dispersion, $\mathcal{H}$ is zero in the vertical degree of freedom and so quantized photon emission does not stimulate vertical emittance growth.

The behavior of the rms energy spread due to quantum fluctuations follows from a similar argument, and so, with the inclusion of the damping terms, the equations of motion for the variances in the three degrees of freedom are:

$$\frac{d\sigma_x^2}{dt} = -\frac{2}{\tau_x}\sigma_x^2 + \tfrac{1}{2}Nf_0\langle\mathcal{H}\rangle\beta_x\frac{\langle w^2\rangle}{E^2}, \qquad (8.39)$$

$$\frac{d\sigma_y^2}{dt} = -\frac{2}{\tau_y}\sigma_y^2, \qquad (8.40)$$

$$\frac{d\sigma_E^2}{dt} = -\frac{2}{\tau_s}\sigma_E^2 + \tfrac{1}{2}Nf_0\langle w^2\rangle. \qquad (8.41)$$

These equations are easily integrated to yield

$$\sigma_x^2(t) = \sigma_x^2(0)e^{-2t/\tau_x} + \tfrac{1}{4}Nf_0\langle\mathcal{H}\rangle\beta_x\tau_x\frac{\langle w^2\rangle}{E^2}(1 - e^{-2t/\tau_x}), \quad (8.42)$$

$$\sigma_y^2(t) = \sigma_y^2(0)e^{-2t/\tau_y}, \qquad (8.43)$$

$$\left(\frac{\sigma_E}{E}\right)^2(t) = \left(\frac{\sigma_E}{E}\right)^2(0)e^{-2t/\tau_s} + \tfrac{1}{4}Nf_0\tau_s\frac{\langle w^2\rangle}{E^2}(1 - e^{-2t/\tau_s}). \qquad (8.44)$$

We see that within a few radiation damping times (assuming that all three degrees of freedom are damped) equilibrium transverse emittances and energy spread are reached:

$$\epsilon_x \equiv \gamma\sigma_x^2/\beta_x \to \frac{1}{2}\left(\frac{\langle\mathcal{H}\rangle}{1-\mathcal{D}}\right)\frac{\langle w^2\rangle}{mc^2\langle w\rangle}, \qquad (8.45)$$

$$\epsilon_y \equiv \gamma\sigma_y^2/\beta_y \to 0, \qquad (8.46)$$

$$\sigma_E/E \to \left[\frac{1}{2}\left(\frac{1}{2+\mathcal{D}}\right)\frac{\langle w^2\rangle}{E\langle w\rangle}\right]^{1/2}. \qquad (8.47)$$

Note it is common practice to quote emittances using $F = 15\%$ in Table 3.1 when discussing electron storage rings. The resulting equilibrium distributions will be Gaussian, as anticipated by our discussion of the central limit theorem in Chapter 7.

While $\langle\mathcal{H}\rangle$ and $\mathcal{D}$ are functions of the accelerator lattice, $\langle w\rangle$ and $\langle w^2\rangle$ are uniquely determined from the photon spectrum generated by synchrotron

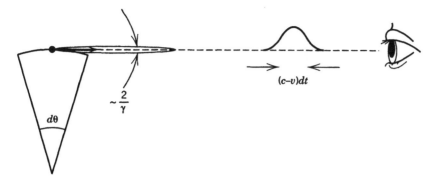

Figure 8.3. An observer sees a pulse of synchrotron radiation of time duration dt' as the cone of radiation sweeps out an angle $d\theta \approx 2/\gamma$.

radiation. A characteristic photon energy can be estimated as follows. Consider a highly relativistic charged particle traveling in a circular trajectory. We know that the radiated energy is concentrated in a cone of angular extent approximately given by $\pm 1/\gamma = mc^2/E$, as shown in Figure 8.3. An observer will see a pulse of radiation which lasts for a time on the order of

$$dt' = \frac{(c-v)\,dt}{c} = \left(1 - \frac{v}{c}\right)dt \approx \frac{dt}{2\gamma^2} = \frac{1}{2\gamma^2}\frac{d\theta}{\omega_0} = \frac{1}{\gamma^3\omega_0}, \quad (8.48)$$

where ω_0 is the instantaneous angular frequency of the circular motion. From Fourier analysis we know that the spectrum of such a pulse will contain frequencies up to about $f_{max} \approx \pi\gamma^3\omega_0$. To see this, consider the Fourier coefficient

$$a_n = \frac{1}{\pi}\int_0^{2\pi} f(z)\cos nz\,dz \quad (8.49)$$

of the function $f(z)$. If $f(z)$ is a pulse of unit height and duration τ which repeats after a period τ_0, then

$$a_n = \frac{1}{\pi}\int_0^{2\pi\tau/\tau_0}\cos nz\,dz = \frac{1}{n\pi}\sin\frac{2\pi n\tau}{\tau_0}$$

$$= \frac{2\tau}{\tau_0}\frac{\sin(2\pi n\tau/\tau_0)}{2\pi n\tau/\tau_0}. \quad (8.50)$$

The coefficients which contribute to the Fourier series will cut off at about $n \approx \tau_0/2\tau$. For our case, $n \approx 2\pi/2\omega_0\,dt' = \pi\gamma^3$. Thus, the maximum pho-

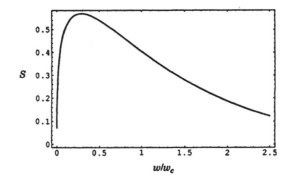

Figure 8.4. Synchrotron radiation power spectrum. The function S is defined in the text.

ton energy should be on the order of

$$w_t \equiv \hbar n \omega_0 = \pi \gamma^3 \hbar \omega_0. \tag{8.51}$$

We will not derive the actual power spectrum. It is given by

$$\frac{dP(w)}{dw} = \frac{P}{w_c/\hbar} S(w/w_c) \tag{8.52}$$

$$S(u) \equiv \frac{9\sqrt{3}}{8\pi} u \int_u^\infty K_{5/3}(v) \, dv \tag{8.53}$$

$$w_c \equiv \tfrac{3}{2}\gamma^3 \hbar \omega_0, \tag{8.54}$$

where K is a modified Bessel function, and w_c is termed the *critical energy*.[2] The function S is shown in Figure 8.4.

In terms of the critical energy, the mean and variance of the distribution are given by

$$\langle w \rangle = \frac{8}{15\sqrt{3}} w_c, \tag{8.55}$$

$$\langle w^2 \rangle = \frac{11}{27} w_c^2. \tag{8.56}$$

[2]See, for example, J. D. Jackson, *Classical Electrodynamics*, Wiley, New York, 1975, and M. Sands, "The Physics of Electron Storage Rings—An Introduction," in *Physics with Intersecting Storage Rings*, ed. B. Touschek, Academic Press, New York, 1971.

We may therefore rewrite our equilibrium conditions as

$$\epsilon_x = \frac{55\sqrt{3}}{2^4 3^2}\left(\frac{\langle \mathscr{H} \rangle}{1 - \mathscr{D}}\right)\frac{w_c}{mc^2}, \tag{8.57}$$

$$\epsilon_y = 0, \tag{8.58}$$

$$\frac{\sigma_E}{E} = \left[\frac{55\sqrt{3}}{2^4 3^2}\left(\frac{1}{2 + \mathscr{D}}\right)\frac{w_c}{\gamma mc^2}\right]^{1/2}. \tag{8.59}$$

For a synchrotron with constant bending radius, the critical energy may be written as

$$w_c = \frac{9}{8\pi}\frac{\hbar c}{r_0}\left(\frac{\rho}{R}\right)\frac{U_0}{E}. \tag{8.60}$$

For typical designs in the $E = 10$ GeV range, U_0 is in the neighborhood of 10 MeV. To allow for straight sections, let's put $\rho/R \approx \frac{3}{4}$. Then $w_c = 19$ keV, that is, in the hard x-ray part of the spectrum. Let $\mathscr{D} \approx 0$ (separated function lattice). If the dispersion function is on the order of 2 m and the amplitude function is on the order of 40 m in the bending regions, then $\langle \mathscr{H} \rangle \approx$ $(2 \text{ m})^2/40 \text{ m} = 0.10$ m. So the equilibrium emittance and energy spread for our example would be $\epsilon_x = 2500$ mm mrad and $\sigma_E/E = 0.8 \times 10^{-3}$. The contribution to the horizontal beam size from the transverse emittance would be $\sigma_x = (\beta_x\epsilon_x/\gamma)^{1/2} = 2.2$ mm. The total horizontal beam size is $\sigma = (\beta_x\epsilon_x/\gamma + D^2 \sigma_E^2/E^2)^{1/2} = 2.7$ mm.

Our idealized results imply that the vertical beam size is zero. In reality, some portion of the transverse emittance will be coupled into the vertical degree of freedom, as will some small part of the horizontal dispersion. Though the vertical beam size will not be zero, for a corrected lattice it will be an order of magnitude or more smaller than the horizontal beam size. As a result, the words "ribbon beam" are often applied to describe the transverse bunch cross section.

PROBLEMS

1. Calculate the radiation per turn lost in synchrotron radiation by
   (a) the Cornell 10 GeV electron synchrotron, whose radius of curvature is 0.1 km, and
   (b) a 5 TeV electron synchrotron built around the earth's equator.
   In each also estimate the bend field.

2. Suppose that a 20 TeV proton storage ring has a circumference of 87 km, a bending radius of 10 km, and a stored current of 70 mA. Calculate the

power going into synchrotron radiation. If that power must be removed from the superconducting magnets, at a temperature of 4 K by refrigerators operating at 20% of ideal Carnot efficiency, estimate the refrigeration power demand.

3. For a separated function synchrotron, oscillations in all three degrees of freedom are damped. Express their time constants in terms of τ_0. Suppose the Fermilab Main Ring were to be used as an electron accelerator. Evaluate the time constants at 20 GeV. (For the Main Ring, $R = 1$ km and $\rho = 0.75$ km.)

4. In a combined function alternating gradient ring, all three degrees of freedom do not damp; in particular, radial betatron oscillations are generally antidamped. Show that this is so for the simple combined function ring of Problem 5 of Chapter 3.

5. In a separated function electron storage ring, show that the (un-normalized) horizontal emittance and the variance of the fractional momentum spread are both proportional to the square of the energy. Evaluate the constants for the Fermilab Main Ring under conditions of Problem 3 above.

6. In an *undulator* electrons traverse a series of magnets, producing alternating up and down fields. The integrated field through this device is zero, so the orbit suffers no net deflection. The angular deviation within a given magnet is within the $1/\gamma$ cone of the synchrotron radiation, so coherence is maintained in the radiation from one magnet to the next. If the pattern of up and down fields has a period length L, show that the synchrotron radiation will have a characteristic wavelength $L/(2\gamma^2)$. This estimate can be made by a variant of the argument used to obtain w_t in the text.

7. In this chapter, the power radiated by a bunch containing n particles is n times the power radiated by a single particle, whereas the factor would be n^2 if the radiation were coherent. Justify the choice of a factor of n.

8. The synchrotron radiation power spectrum, Equation 8.52, is calculated on the basis of classical electrodynamics. We could expect this result to be valid provided the critical energy, w_c, is small compared with the energy of the particle. Some of the challenging parameter sets that have been put forward for linear electron colliders imply operation in a quite different regime. For order-of-magnitude purposes, suppose n particles of (total) energy E are uniformly distributed in a cylinder of radius r and length L. A particle traveling in the opposite direction with the same energy intercepts the bunch at radius r; it will emit synchrotron radiation (called "beamstrahlung") due to the electromagnetic fields of the bunch. Estimate the ratio of the critical energy to the total energy based on the classical picture for $\gamma = 10^7, n = 10^9, r = 10^{-9}$ m, and L $= 10^{-6}$ m.

Tables of Accelerator Parameters

A number of problems use some of the Fermilab proton accelerators as examples. The following tables may be useful in the event that the statements of the problems are confusing or incomplete. The parameters for the Main Ring relate to its present role as an injector for the Tevatron, so the maximum energy is only 150 GeV rather that the 400 GeV of the pre-Tevatron era.

Table A.1. Booster synchrotron.

Circumference	$2\pi \times 74.47$ meters
Injection energy	200 MeV (kinetic)
Peak energy	8 GeV (kinetic)
Cycle time	$\frac{1}{15}$ sec
Harmonic number, h	84
Transition gamma	5.45
Maximum RF voltage	0.86 MV
Longitudinal emittance	0.25 eV sec
Horizontal β_{max}	33.7 meters
Vertical β_{max}	20.5 meters
Maximum dispersion	3.2 meters
Tune $\nu_x \approx \nu_y$	6.7
Transverse emittance[a]	8π mm mrad
Bend magnet length	2.9 meters
Standard half-cell length	19.76 meters
Bend magnets per cell	4
Bend magnet total	96
Typical bunch intensity	3×10^{10}
Phase advance per cell	96 deg
Cell type	FOFDOOD

[a]Normalized, 95%.

Table A.2. Main Ring as Tevatron injector.

Circumference	$2\pi \times 1000$ meters
Injection energy	8 GeV (kinetic)
Peak energy	150 GeV
Cycle time	≥ 2.6 sec
Harmonic number, h	1113
Transition gamma	18.7
Maximum RF voltage	3 MV
Longitudinal emittance	0.25 eV sec
β_{max} in insertion	225 meters
β_{max} in cells	100 meters
Maximum dispersion	6 meters
Tune $\nu_x \approx \nu_y$	19.4
Transverse emittance[a]	12π mm mrad
Bend magnet length	6.1 meters
Standard half-cell length	29.7 meters
Bend magnets per cell	8
Bend magnet total	774
Typical bunch intensity	2×10^{10}
Phase advance per cell	68 deg
Cell type	FODO

[a] Normalized, 95%

Table A.3. Tevatron — collider mode.

Circumference	$2\pi \times 1000$ meters
Injection energy	150 GeV
Peak energy	900 GeV
Acceleration period	52 sec
Harmonic number, h	1113
Transition gamma	18.7
Maximum RF voltage	1.4 MV
Longitudinal emittance	3 eV sec
β_{max} in insertion	900 meters
β_{max} in cells	100 meters
β^* at collision point	0.5 meter
Maximum dispersion	12 meters
Tune $\nu_x \approx \nu_y$	19.4
Transverse emittance[a]	24π mm mrad
Bend magnet length	6.1 meters
Standard half-cell length	29.7 meters
Bend magnets per cell	8
Bend magnet total	774
Typical bunch intensity:	
protons	1×10^{11}
antiprotons	5×10^{10}
Phase advance per cell	68 deg
Cell type	FODO

[a]Normalized, 95%.

Bibliography

The first five references are the proceedings of the U. S. Particle Accelerator School. The level of the articles covers the whole spectrum from introductory to advanced. Many of the treatments of specialized topics include extensive bibliographies. The present book is an outgrowth of an article in the fifth of these references.

Courant and Snyder is the classic paper on the alternating gradient synchrotron, and is "must" reading for a serious student of accelerator physics. Though not published until 1958, the bulk of the work was done in late 1952 and early 1953.

The first CERN publication listed, though it is a compilation of articles by many authors, has the character of an introductory text that can be read from cover to cover. The other CERN books are proceedings from CERN schools equivalent to the U.S. summer schools.

The proceedings of the national and international conferences are a rich source of information on just about any aspect of accelerator science and technology.

1. *Physics of High Energy Particle Accelerators* (Fermilab Summer School 1981), ed. R. A. Carrigan, F. R. Huson, and M. Month, AIP Conf. Proc. 87, AIP, New York, 1982.
2. *Physics of High Energy Particle Accelerators* (SLAC Summer School 1982), ed. M. Month, AIP Conf. Proc. 105, AIP, New York, 1983.
3. *Physics of High Energy Particle Accelerators* (BNL-SUNY Summer School 1983), ed. M. Month, P. F. Dahl, and M. Dienes, AIP Conf. Proc. 127, AIP, New York, 1985.
4. *Physics of Particle Accelerators* (SLAC Summer School 1985, Fermilab Summer School 1984), ed. M. Month and M. Dienes, AIP Conf. Proc. 153, AIP, New York, 1987.
5. *Physics of Particle Accelerators* (Fermilab Summer School 1987, Cornell Summer School 1988), ed. M. Month and M. Dienes, AIP Conf. Proc. 184, AIP, New York, 1989.

6. E. D. Courant and H. S. Snyder, "Theory of the Alternating Gradient Synchrotron," Ann. Phys. **3**, 1 (1958).

7. *Theoretical Aspects of the Behaviour of Beams in Accelerators and Storage Rings* (Proc. First Course of International School of Particle Accelerators of "Etore Majorana" Centre for Scientific Culture), ed. M. H. Blewett, CERN 77-13, July 1977.

8. *Antiprotons for Colliding Beam Facilities* (Proc. CERN Accelerator School 1983), ed. P. Bryant and S. Newman, 1984.

9. *General Accelerator Physics* (Proc. CERN Accelerator School 1984), ed. P. Bryant and S. Turner, 1985.

10. *Advanced Accelerator Physics* (Proc. CERN Accelerator School 1985), ed. S. Turner, 1987.

11. *Frontiers of Particle Beams; Observation, Diagnosis, and Correction*, ed. M. Month and S. Turner, Springer-Verlag, Berlin, 1989.

12. The proceedings of IEEE Particle Accelerator Conferences, held every two years (1991, 1989, 1987, . . .), published by The Institute of Electrical and Electronics Engineers, New York.

13. The proceedings of International Particle Accelerator Conferences, held every three years (1989, 1986, 1983, . . .), publishers vary—most recently, Particle Accelerators, Vols. 30, 31 (1990), Gordon and Breach Science Publishers, New York.

14. *Linear Accelerators*, ed. P. M. Lapostolle and A. L. Septier, North-Holland, Amsterdam, 1970.

15. M. Sands, *The Physics of Electron Storage Rings—An Introduction*, SLAC publication SLAC-121, UC-28 (ACC), 1970.

Index

Printed in the USA/Agawam, MA
November 13, 2020

764444.002